tp1.2

A continuing inquiry
into the
Foundations of
the Science of Physics:

Vector Algebra

JRBreton

tp1.2
A continuing inquiry into
the Foundations of the Science of Physics:
Vector Algebra

by JRBreton

Published by:
The Foundation for Theoretical Physics
3 Apple Tree Lane
Walpole, MA 02081−2301

Web address: FoundationForTheoreticalPhysics.org
email address: theoretical.physics.books@gmail.com

Copies of this book and other offerings of the Foundation may be obtained by order from the Foundation's address, preferably online from tp.vendevor.com, the Foundation's website store and also from amazon.com, Barnes&Noble, and similar sites.

All rights reserved. No part of this book may be reproduced in any form by any means, electronic or mechanical, including photocopying, recording, or any information storage or retrieval system, without written permission from the author, except for the inclusion of brief quotations in a review.

Copyright © 2020 by JRBreton

ISBN, print ed.   978−0−9844299−8−1

First printing 2020

Printed in the United States of America

Library of Congress Control Number: 2020915795

Table of Contents

The Next Day..5
Vectors..8
    Addition in the axiomatic set of vectors...21
        Peaches...24
    Subtraction in the axiomatic set of vectors..38
        Ark of the Covenant...41
    Multiplication in the Set of Vectors...46
        Inner (Dot) Product..47
        Cross Product..51
        Outer Product..53
            Sums of inner products...55
    Combinations of Multiplications...59
        The scalar triple product..59
The Origin..71
        Sums of cross products...80
        Sums of outer products..83
        The vector triple product..84
        Other Identities..100
        TABLE OF VECTORIAL IDENTITIES...109
Solutions of vector algebraic equations...114
    Of equations with inner products...114
    Of equations with cross products...122
    Of equations with outer products..126
    TABLE OF SOLUTIONS TO VECTORIAL EQUATIONS.......126
Solutions to Matrix equations..127
    Solutions for the Matrix..152
Transformation into Theoretical Physics...159
    References..160
        Material/Spatial References...162
    Observations..163
        The Principle of Non-Collocation..164
    Vectorial Multiplication in Theoretical Physics...................................167
    Some Reflections on the Science of Physics..171

The Next Day

The morning of the next day found the three friends, Newton, Einstein, and Breton, well breakfasted, in their clubhouse, seated by the fire in their comfortable Windsors, each harboring a tempered eagerness to continue their investigation of Theoretical Physics. The smell of Autumn danced again with the sound of a log crackling in the fireplace. The room was no less cozy and peaceful; warm but not hot; lighted, but not bright.

Newton, who loved to summarize and recapitulate as well as construct tables, began the conversation with an attempt to summarize the conclusions of the previous day. "We discussed so many subjects yesterday, let us start by summarizing yesterday's conclusions. We came to see how a science like Physics differs from a technology like Surveying."

Breton, interrupting: "Technology is useful, and when no longer useful, discarded. It willingly sacrifices truth for utility. Science is not necessarily useful but will not compromise truth and so is permanent."

Einstein, adding in his friendliest tone: "Technology relies on measurement, but science does not."

Newton: "And many more differences which you will remember from my tables of yesterday, but sciences differ from each other too. The science of Physics differs from the science of Mathematics."

Einstein, happily contributing: "They are both true and permanent, but differ in what they are true to."

Breton, happily contributing also: "Mathematics is true to its axioms while Physics must be true to some aspect of reality."

Newton, leading contentedly: "We accepted, after some debate the following definition of Physics:
> **Physics** is the study of reality observable as extended, moving, or forcing.

Breton: "So we concluded that the symbols and ideas of Mathematics were inappropriate for the science of Physics because they easily lead to ambiguities and false conclusions."

Einstein, in a somewhat less friendly tone,: "Physics without Mathematics, that is hard to swallow since everywhere Physics is explained in terms of the symbols of Mathematics."

Breton, accepting the implied challenge: "Yes, we saw how treacherous the abuse of language and symbols becomes for anyone seeking the truth in any science. Ordinary language abounds in ambiguities; we noted the emergence of dictionaries to help us communicate.

   Since truth in general and scientific truth in particular cannot tolerate contradictions, sciences are forced to construct special dictionaries to avoid ambiguities in their disciplines. Mathematics has its dictionary; Physics should also have its own, distinct from the one for Mathematics. We call the dictionary for the science of Physics by the name Theoretical Physics. It, not Mathematics, is the proper language for Physics."

Einstein, doggedly: "But for all that, the language of Mathematics does seem appropriate for the study of Physics."

Breton: "Mathematics has great appeal because of its simple, logical structure, a quality which should also characterize an appropriate language for Physics. Nevertheless the ideas of Mathematics are not the ideas of Theoretical Physics, even when they to use the same symbols. In Mathematics a symbol would refer to the mathematical dictionary, while the same symbol in Theoretical Physics would refer to a different dictionary. A grave confusion results from using the wrong dictionary."

Newton, placidly: "We illustrated all this. For instance, the proposition
$$2+2=4$$
is true enough as a mathematical statement, but ambiguous or even false as a physical statement."

Breton, continuing: "So we embarked on a great adventure to discover how mathematical ideas and propositions can be transformed into ideas and propositions suitable for Theoretical Physics. The outlines of the adventure are clear enough: to transform any mathematical idea it must first be constrained, and when so constrained can then be elaborated into a panoply of related ideas."

Einstein: "So we examined some mathematical ideas with a view to their transformation into Theoretical Physics."

Newton: "We started with the positive integers, a subject I could not imagine held such amazing profundities. From there it was more amazement with the negative integers and then even more with quotient numbers. Quotient numbers, we discovered, harbor a topology, from which the mathematical ideas of limits germinate. From there we examined the amazing world of functions, and ideas of continuity, derivatives and integrals."

Breton: "You've omitted so much."

Newton: "True enough: topics like look-alike functions, and step functions, and many others besides. When I reflect on our conversation yesterday I stand amazed at the many and amazing topics we wrestled with. A brief summary just omits too much. Yesterday's conversation should be made into a book!"

Breton: "What would be its title?"

Newton: "Why don't *you* propose a title?"

Breton: "Let's title it 'tp1.1'. The title would stand for theoretical physics 1.1. The 1.1 would indicate more to come."

Einstein: "Who, except us, would know what tp1.1 means?"

Breton: "We could give it a subtitle like 'an inquiry into the foundations of the science of Physics.'"

Einstein: "Better, but still obscure."

Breton: "Would *you* like to try your hand at a title?"

Einstein, after a pause: "No. The book might appeal to adventurous and inquiring minds and surely discourage shallow, superficial surfing. Let the title stand."

Newton: "We can refer to the book *tp1.1* for any of the many topics we have not summarized."

Einstein, looking to assert his standing in the discussion: "But Breton points out that none of this amazing world of Mathematics is Theoretical Physics. We ended the day by showing how these mathematical ideas can be transformed into Theoretical Physics. We would first give any mathematical idea a physical label. Physical labels are all reducible to three elementary ones: for extension (L), for motion (V), and for force (F). Mathematical expressions, being unlabeled, may be combined in ways that are not allowable for Theoretical Physics. So it became apparent that, although an identical symbol might be used, a number in Mathematics is a different idea from a number in Theoretical Physics."

Newton, continuing: "Expressions in Theoretical Physics must follow the Rules for Labels. The Rules show how new ideas for Theoretical Physics can be created from the elementary ones."

Einstein, still looking to lead: "In addition to labeling, we saw how the ideas of Theoretical Physics must be referenced either materially or spatially."

Breton: "And how the word 'set' can said of material things as well as mathematical ideas which led to the idea of a particle, the properties of material things, and the constraints of resolution."

Einstein: "We would do well to deepen our conversation about these topics since they promise application to the science of Physics."

Newton: "But Breton suggested that today we look into the subject of location."

Breton, looking to smooth the rising contention between his friends: "Thank you Newton. We observe physical objects located here and there. Yet very little of what we discussed yesterday faced this aspect of physical reality. Mathematics provides an interesting structure called *vectors* which may be suitable for transformation into Theoretical Physics. I suspect we will deepen our knowledge of yesterday's topics by seeing them in this new perspective. Will you trust me, Einstein?"

Einstein: "I trust an old friend, or rather, I agree based on our fast friendship, but be forewarned I will be quick to object if that trust is violated."

Breton: "Thank you, Einstein. I will do my best not to violate your trust.

Vectors

Einstein, always striving to lead: "Start by giving us a definition."

Breton: "That is difficult because a vector is an elemental idea. There aren't many antecedents upon which I can build a definition. For instance, if I defined a vector to be an element in the axiomatic set of vectors you would say immediately that that defines nothing."

Einstein, with not a little sarcasm: "Without a definition we don't know what we are talking about."

Breton: "Agreed. What elemental experience can I refer to for a start?"

Newton: "you noted yesterday that of the many topics we discussed, nothing touched location. Yet Physics deals with extension, motion and force, all of which imply a location at which an object can be observed as extended, moving, or forcing. So let me suggest we take location as a given upon which to build a definition."

Breton: "Let's try this definition."

> **Definition** (vector)
> Given
>   the location of an object
> then
>
>   a **vector** is an idea which specifies its location.
>   end of definition

Einstein: "The definition doesn't say much. A vector then is just another name for location."

Newton: "Breton, your definition has a physical flavor to it. We are embarked on an adventure to convert mathematical ideas into those suitable for Theoretical Physics. Let's start with some purely mathematical ideas."

Einstein, trying to be helpful now that Breton had been cornered: "How did we approach positive integers? We did not define numbers, we simply enumerated them, and declared they were subject to a plus operator."

Newton: "Or alternatively we declared the positive integers to be the result of an indefinite application of the plus operator on a number called one."

Breton: "So, we should be looking for axioms, rather than definitions?"

Einstein: "What's the difference?

Breton: "Axioms are fundamental statements upon which a logical structure can be erected. Like rules for a game, they need to be simply accepted. If the axioms are changed a different structure will emerge. Think of Euclid's axioms for geometry. They form a basis for a plane geometry. Change the axioms, a new geometry will appear.
   Definitions are built upon the axioms. They use the accepted axioms, including their terms, as a root vocabulary."

Newton: "How does this fit in with location?"

Breton: "Location is an attribute of an object. If the object is a material one, location is a physical attribute, not an idea at all. A vector is an *idea* which hopefully can be transformed to describe a location. To provide all possible descriptions for locations, we create a set ideas of all possible lengths and all possible directions."

Einstein, looking to derail a coming argument: "But what if the object is a mathematical idea like a triangle?"

Breton placidly: "Before triangles, we should first discuss angles."

Einstein: "And before angles, lines and points."

Breton: "So we have entered into a discussion of geometry, a vast subject which may only be related to our goal tangentially."

Newton: "My illustrious ancestor loved geometry. Let us honor the great man by stating at least the foundations. Geometry consists of lines which may intersect at points."

Breton, trying to angle a return towards the main goal. "The point at which two lines intersect is called a **vertex**. At the vertex four **angles** are formed between the lines. There are many kinds of angles; it will be worth our while to define them and then consider how they apply to triangles."

Einstein, happily interrupting,: "And how do you measure angles?"

Breton, always looking to accommodate: "You bring up another good point. Angles, indeed, can be measured because they have parts. As a mathematical idea, an angle is complex. We started with two lines which intersect. The intersection, called the **vertex**, can form the center of a circle. Further, from the vertex we can truncate one of the lines finitely and let it be the radius of a circle. An **angle** is defined and measured in this complex of lines, vertex and circle. To measure the angle, note that two lines, intersecting the circle, define an arc of the circle. The ratio of the length of the arc compared to the circumference of the circle is used to measure the angle."

With that Breton quickly sketched this illustration.

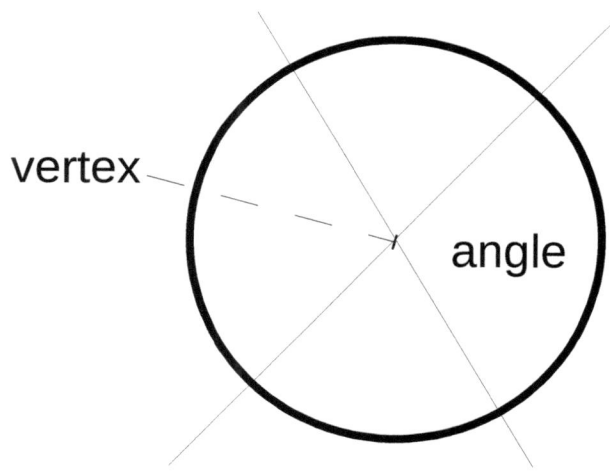

Newton: "So different measures result from the measurement of the circumference."

Breton: "Exactly. Two of the most common are called **measurement in degrees** and **measurement in radians**. For measurement in degrees the circumference is divided into 360 equal parts. The arc of the angle will then be measured as so many of these degrees. For measurement in radians the circumference is divided into the number of radii which will fit into it. That number is 2*pi. The arc of the angle will then be measured as so many of these radii, called **radians**."

Einstein: "So the actual measurement is accomplished in terms of arbitrary units."

Breton: "Not arbitrary. The measurement assumes a reference which must be stated (but often merely assumed) when the unit is declared.
If all this is clear, let us return to the definitions of different angles.
...An **acute** angle is an angle less than 90 degrees (pi/2);
...a **right** angle is an angle equal to 90 degrees (pi/2);
...an **obtuse** angle is an angle greater than 90 degrees (pi/2), but less than 180 degrees (pi);
...a **reflex** angle is an angle greater than 180 degrees (pi).

Two angles are sometimes paired. The following definitions are often helpful.
Two angles are called **conjugates** if their sum equals 360 degrees (2*pi).
Two angles are called **supplementary** if their sum equals 180 degrees (pi).
Two angles are called **complimentary** if their sum equals 90 degrees (pi/2)."

Turning to Newton, Breton asked: "Newton, would you please make a table of these definitions for us?"

Quickly responding, Newton quickly produced the following table.

Angles	
Type	Definition
acute	less than pi/2
right	equal to pi/2
obtuse	greater than pi/2
reflex	greater than pi
conjugate	sum equals 2*pi
supplementary	sum equals pi
complimentary	sum equals pi/2

Einstein: "Then a **triangle** is a mathematical structure of lines forming three angles."

Breton, returning to the main track gleefully: "Then the *location* of one vertex can be referred to a second vertex. In a similar way the location of a physical object must be referred to some observer. You bring up some good points Einstein.

In surveying, mathematical triangles play no unimportant role. Triangles are not numbers. Would it be worthwhile to begin our study of location and vectors with triangles?"

Einstein, needling Breton: "Since we are searching for foundations, angles would be a better choice. Don't you agree angles are more basic than triangles?"

Breton, humbly conceding,: "Agreed."

Newton: "And triangles? I want to discuss triangles further."

Breton: "A triangle is a mathematical object with only three angles. It will then have only three vertices, and then each vertex will share two lines. These shared lines are called **sides** of the triangle, and they number three also.

Usually the triangle is a planar figure, but not necessarily so. Even when restricted to a plane, the plane need not be a Euclidean plane. A large variety of triangles may be defined since the three angles need not all be the same.
...An **oblique** triangle is one all of whose angles are acute;
...a **right** triangle is one which has one right angle;
...an **equilateral** triangle is one all of whose angles are equal;
...an **isosceles** triangle is one with two equal angles;
...a **scalene** triangle is one none of whose angles are equal."

Newton, anticipating a request quickly produced the following table without being asked.

Triangles	
Type	**Definition**
oblique	all angles acute
right	one right angle
equilateral	all angles equal
isosceles	two angles equal
scalene	no angles equal

Breton: "Since we are talking about triangles, let me produce a diagram which indicates generally the angles and sides of a triangle and the convention to use in referring to them. The simple numbers reference the angles. The symbols l1, l2, l3 reference the sides opposite the numbered angles.

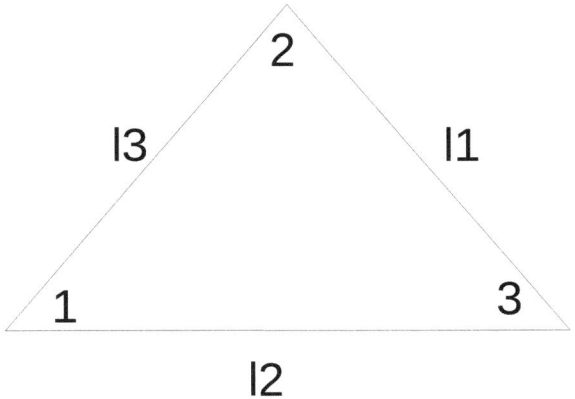

With reference to oblique triangles the following two 'laws' of triangles can be stated.

Sine Law: sin(angle(1))/length(l1)
          = sin(angle(2))/length(l2)
          = sin(angle(3))/length(l3)
Law: of Projections:
length(l1) = length(l2)*cos(angle(3))+ length(l3)*cos(angle(2))

Einstein: "You haven't defined what sin(angle) or cos(angle) means, much less proven the laws!"

Breton: "Little gets by you Einstein. You're right. These ideas are defined for any angles, but come from a couple of special definitions associated with right triangles. First let me sketch a right triangle.

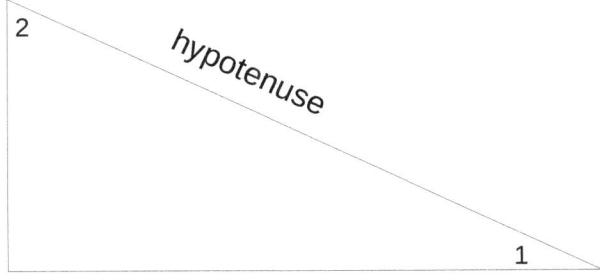

The side opposite the right angle is called the **hypotenuse**. The other two sides are **orthogonal** to each other.

The following two definitions should be remembered.
$$\sin(\text{angle}) \equiv \text{length of the side opposite the angle} / \text{length of the hypotenuse}$$
$$\cos(\text{angle}) \equiv \text{length of the side nearest the angle} / \text{length of the hypotenuse}$$
As you can see from the sketch
$$\sin(\text{angle1}) = \cos(\text{angle2})$$
and
$$\sin(\text{angle2}) = \cos(\text{angle1}).\text{"}$$
Let me defer the proof of the laws for the moment.

Einstein, taking charge of the discussion again: "Just give us a definition of a vector."

Breton: "Rather let me give you the *axioms* of an axiomatic set of vectors which we may symbolized as **V**. The set is populated by elements called vectors symbolized by **v**. This axiomatic set of vectors presupposes Q, the set of quotient numbers with its algebra and topology. The axioms also presuppose two operations, vector addition (symbolized by +) and multiplication of a vector by quotient numbers called **scalar** multiplication (symbolized by ∗). The vectors obey the following axioms:

  v1 + (**v2** + **v3**) = (**v1** + **v2**) + **v3**:
  v1 + **v2** = **v2** + **v1**
  q1∗(q2•**v1**) = (q1∗q2)•**v1**
  q1∗(**v1** + **v2**) = q1•**v1** + q1•**v2**
  (q1+q2)•**v1** = q1•**v1** + q2•**v1**
  1∗**v** = **v**, for any vector in the axiomatic set of vectors.

Also, there exists a zero vector in the axiomatic set of vectors symbolized by **0** such that
$$\mathbf{v} + \mathbf{0} = \mathbf{v}$$
for any vector, and for every vector **v** in the axiomatic set of vectors there exists a vector −**v** such that
$$\mathbf{v} + (-\mathbf{v}) = \mathbf{0}$$
Does this help?"

Einstein: "Not much, if at all."

Newton: "Why not call this axiomatic set of vectors a space?"

Einstein: "It can be used to show the relative positions of objects."

Breton, mildly correcting: "Their locations. Is space something physical or is it a mathematical idea?"

Einstein: "Physical!"

Newton: "Mathematical!"

Breton: "Your answers show that this question should be addressed. It seems to me that some uses of the word 'space' are physical or quasi-physical, and others mathematical. Let me list some current uses of the word:

outer space, as the universe beyond earth's atmosphere;
as a gap between written characters. ASCII code 32;
personal space in human relationships;
a square in a board game;
a business term to describe a competitive environment;
a solution space, candidates for solutions of equations;
mental space in cognitive science;
a vacuum;
some buildings;
address space in computers;
cyberspace;
white-space as allocated but locally unused radio frequencies."

Newton: "Enough. The word 'space' can be used in science and engineering, but also in fiction, music, art, law and many, many other contexts."

Breton: "It appears the word 'space' can be stretched indefinitely. If used as a synonym for void, then it seems to me poorly adapted to what I have called an axiomatic set of vectors."

Newton: "Remember the ambiguities of language we discussed in tp1.1? Aren't we slipping into such a morass when we use the word space?"

Breton: "The word 'space' may be useful to us later, but only after it has received a rigorous definition. Remember the story of the mountain hiker?"

Einstein: "Let's just use the word 'set' which is less ambiguous and can refer to mathematical objects unequivocally."

Breton: "Fine. So we can look to examine vector sets. But let us reflect on the intellectual path we have covered. First we thought to define the set **V**.
$$\mathbf{V} \equiv \{\mathbf{v}\}$$
as a set of objects. Einstein rightly remarked this definition said little. Next, we specified further
$$\mathbf{V} \equiv \{\{\mathbf{v}\}, Q, +, *\}$$
where Q is our algebra of quotient numbers, + a new kind of addition, and * a new kind of multiplication. While we have defined Q previously in tp1.1, the two operators remain unspecified."

Newton, in a moment of discovery: "Look! The axioms for the axiomatic set of vectors have extended Q. If we take the vectors to be the partitions of the quotient numbers, and plus and multiply as defined for Q, then Q is itself an axiomatic set of vectors."

Breton: "We are well started then. The axiomatic set of vectors will be a set rooted in Q, associated with it by scalar multiplication, but possibly developed far beyond Q. So if $\mathbf{v}$ is a vector and q1 is a quotient number, then q1*$\mathbf{v}$ is also a vector."

Einstein, objecting: "Hold it there! This is a strange multiplication different from any of the others we seen."

Breton: "True enough. Very little gets by you, Einstein. This is yet another kind of multiplication, a multiplication of a different color."

Einstein: "Why call it multiplication at all?"

Breton: "Remember our rules for symbols? We might have written
$$q1 + \mathbf{v1}$$
but that would violate our rules for labels since q1 and **v1** should then have the same label. Since we are looking to transform mathematical ideas into Theoretical Physics we would do well to rule out such combinations.

But according to the rules for labels, the symbol
$$q1 * \mathbf{v1}$$
would be acceptable."

Einstein: "But why call it multiplication at all?"

Breton: "Mathematicians call it **scalar** multiplication, a special kind declared for vector sets. The word scalar is used because operations like
$$q1 * q2 \bullet \mathbf{v1}$$
are allowed. This kind of operation 'scales' the vector by making it longer of shorter."

Einstein: "Now you're mixing the multiplications. The first applies only to Q; the second only to the axiomatic set of vectors. They must be different operators."

Breton: "Right again. They are *different* operators and must be seen as such. For his reason I have emboldened the second symbol for multiplication. I ask you to tolerate this admitted confusion for simplicity's sake. Scalar multiplication is stipulated by the axioms of an axiomatic set of vectors."

Einstein, thinking to corner Breton again: "This could get confusing."

Breton, naively: "It gets worse. Mathematicians also endow the vector set with some algebraic features. So the axiomatic set of vectors has its own plus operator such that if **v1** and **v2** are vectors then
$$\mathbf{v1} + \mathbf{v2}$$
is also a vector."

Einstein, now looking to scope out the problem: "We know about Q, and, as Newton has observed, by itself it can be considered a vector set. On the other hand, vector sets other than Q may exist—a mathematical universe whose extent is hard to imagine."

Breton, continuing unabashed: "So from this mathematical universe of vector sets, we have to pick one which can be made suitable for Theoretical Physics."

Newton: "How?"

Breton: "You have already suggested the right path. We are looking for a vector set which can be useful in describing locations, just simple, mathematical locations."

Einstein, scornfully: "What is a simple, mathematical location?"

Newton: "That's easy enough. In conformity to its axioms, the vector set has a zero vector. A simple mathematical location is the vector related to the zero vector. For instance, my illustrious ancestor, could describe the position of a planet with relation to the sun as a vector relative to the zero vector taken as the sun."

Breton, noting a flashing from Einstein's eyes, pressed on quickly to avoid a confrontation: "So what is needed is a vector which incorporates both a distance and a direction. The distance can be taken as a scalar from the quotient numbers, Q, but what about direction?"

Einstein, stubbornly: "I don't like the word 'distance'. It connotes an idea which relates more to physical reality. Let's use a more mathematical word like 'length'."

Breton, in a conciliatory mood: "Fine. So let every vector in our vector set be composed of a length and a direction."

Newton, helpfully: "Direction can be thought of as points on a sphere centered on the zero vector."

Breton, probing: "So can directions be taken from Q?"

Newton, with a touch of irritation: "No, directions seem to be some kind of vector themselves without length."

Breton: "Or with some implied length. In any case the vectors in our vector set would have a scalar length and a vector direction. The length and the direction could not be added, but they could be symbolized as
$$\mathbf{v} = q * \mathbf{u}$$
where the emboldened characters represent vectors and the unemboldened character represents a quotient number."

The multiplication sign annoyed Einstein. Despite his insistence on dis-ambiguity, Breton was clearly using the symbol ambiguously. Looking to draw out some consequences from the inconsistency, Einstein continued: "Then any vector, say $\mathbf{v}$, in our special vector set could be written as
$$\mathbf{v} = q(\mathbf{v}) * \mathbf{u}(\mathbf{v}).\text{"}$$

Newton, seizing the argument with a certain bustle: "If
$$\mathbf{v} = u(\mathbf{v})$$
then $q(\mathbf{v})$ must equal one!"

Breton, happily concluding: "So direction is a vector whose length is one. We might call such vectors **unit vectors**."

Einstein, somewhat miffed because the argument had not gone as he expected: "I suspect you anticipated all this by symbolizing 'u' for directions since they are unit vectors."

Newton, ignoring Einstein's ignoble suggestion, wrapped up the conclusion: "Then directions are simply unit vectors, one for each point on a unit sphere centered on the zero vector."

Einstein: "Length and direction are measurable. Have we fallen from science into technology? Remember we agreed that technology relies on measurement, a trait that separates technology from science."

Breton, patiently: "Measurements result in numbers; vectors are not numbers. Recall our previous discussion. Theoretical Physics deals with objects that are measurable, because they are extended. Relationships between its ideas can be explained without actually measuring anything, just as relationships between mathematical ideas can be explained without measurements."

Newton, pushing on: "Are vectors ordered?"

Breton: "You ask a good question, Newton. Remember our discussion of quotient numbers? Quotient number are not ordered of themselves, but when partitioned correctly, the partitions themselves emerge as ordered. Something analogous occurs with vectors. The axiomatic set of vectors is not ordered, but certain subsets may be."

Einstein: "Length can be stated in numbers, but what about direction?"

Breton: "The numbers related to length are called the **underlying field** of the axiomatic set of vectors and for Theoretical Physics the set of partitions of quotient numbers, called Q, is sufficient. Many mathematicians, however, prefer to use the real numbers, R, as the underlying field."

Einstein: "I prefer to say 'completed numbers', rather than real numbers."

Breton: "As you will. I mean to emphasize that the underlying field comes with its algebra and topology."

Einstein: "How about direction?"

Breton: "Imagine you are standing in the center of a sphere. Any point on the sphere from your perspective would be a direction."

Newton: "So these imaginings give us some idea of a vector, but one I find little helpful. Let me suggest another approach. Remember how the positive integers were developed? Yesterday, Breton asked me for not one integer, but the whole set of them."

Einstein: "Then I proposed how the whole set could be generated by an algorithm."

Breton: "And one of you, I don't remember which, questioned why we should develop the whole set at all."

Einstein: "I did."

Breton: "We answered that the whole set was developed to answer any question about the operation of the plus operator on any positive integer. So the development could be seen as forming the set of answers to all such questions."

Newton: "Why not try the same with vectors? What is the axiomatic set of vectors good for?"

Breton: "A splendid insight Newton. Let us consider the entire axiomatic set of vectors, covering every length and every direction."

Einstein: "Like the quotient numbers."

Breton: "Something like, but not the same. The directions of vectors refer to a sphere, while the directions of quotient numbers refer to a circle."

Einstein: "So the whole axiomatic set of vectors provide answers about locations."

Newton: "Vectors are growing a little more useful, but not much."

Einstein: "You represent direction as a vector. Isn't direction an angle?"

Breton: "You ask an interesting question. Which is more fundamental: angle or direction?"

Einstein: "Angle, because directions are stated in terms of angles."

Breton: "But angles need a reference. Isn't an angle measured between two directions? So it seems to me that direction is the more fundamental concept. I can point to something to show its direction without any reference to an angle."

Einstein, doggedly: "In any case angle and direction are closely related ideas."

Breton: "Different, but related. Moreover, once a system of orthogonal axes is accepted, any point on the sphere can be located by three angles. If the three angles are called angle1, angle2, and angle3, then the three angles defining a given direction are constrained by the by the formula
$$(\cos(angle1))^2 + (\cos(angle2))^2 + (\cos(angle3))^2 = 1$$
where cos is the trigonometric function called cosine."

Einstein, never an amiable loser, thought to change the direction of the conversation: "Vectors are members of a vector set, so for any two vectors **v1** and **v2**, another vector, **v1+v2**, is a member of the same set. What is **v1+v2**?"

Newton, agreeably, without noticing Einstein's tactic: "Let's also consider scalar multiplication."

Breton, taking up the new challenge: "Scalar multiplication is easy. If **v1** = q1•u(**v1**) is a vector, then
$$\mathbf{v2} \equiv q2 \bullet \mathbf{v1} = q2*q1 \bullet u(\mathbf{v1})$$
So scalar multiplication produces another vector having the same direction as the original vector, but with a scaled length. For this reason this kind of multiplication in the vector set is called scalar multiplication.

Two vectors having the same direction are called **parallel vectors**."

Newton, probing: "Suppose **v2** = (–1)•**v1** are they still parallel?"

Breton: "Good remark, Newton. I need to be more precise. Let me offer the following definition.

Definition (parallel vectors)
    Given
            V, an axiomatic set of vectors Q
      for
            v1 and **v2**, vectors in **V**
      if
            there exists a q1 in Q such that
            v1 = q1•**v2**
    then
            v1 and **v2** are **parallel vectors**.

    Further if q1> 0
        v1 and **v2** are **positively parallel vectors**;
    and if q1< 0
        v1 and **v2** are **negatively parallel vectors**.
                      end of definition

Newton, probing still: "How about **0**?"

Breton: "Arguing from the definition, the zero vector would be parallel to any vector in **V**.

Einstein, making a favorite point and attempting again to control the conversation: "So the zero vector is a special vector! Let's return to vector addition. How about **v1+v2**?"

Addition in the axiomatic set of vectors

Breton, taking up the challenge gingerly: "The vector, **v1+v2**, will have a length and direction, so
$$v1+v2 = q \bullet u;$$
so to define vector addition we have to define q and **u**."

Einstein, with a touch of triumph in his voice: "How?"

Breton, pensively and cautiously: "Let's try with a simple example. Suppose **v1** and **v2** are directions. If so,
$$u(v1) + u(v2) = q \bullet u$$
$$= 2*\cos(\text{angle}/2) \bullet u$$
where angle is the angle between **u(v1)** and **u(v2)**."

Einstein, quickly objecting: "Where did that come from? Why not
$$u(v1) + u(v2) = u(v1+v2)?$$
You specify q but not **u**, while I specify both."

Breton, now aware of Einstein's truculence: "Really? You specify that the vectorial sum of two directions is another direction, but still unspecified.
  But from the axioms of the vector set
$$u(v1)+u(v1) = 2*u(v1)$$
which is no longer a unit vector. So it appears your suggestion implies a contradiction and so cannot be considered an appropriate definition."

Einstein, crushed, but too proud to concede: "Are there contradictions with your suggestion?"

Breton: "Could be. As you pointed out, my suggestion does not specify the direction."

Thinking prudence called for, Einstein thought for a long moment on how best to challenge Breton. Finally, he mused: "I note that you have defined an angle from two directions. We might have defined direction in terms of angles. So which is a more fundamental concept: angle or direction?"

Newton, impulsively: "Angle!"

Einstein, glad to see his question taking root: "I say direction!"

Breton, always looking to reconcile his two friends: "Besides simple assertions, how can we come to the truth of the matter?"

Newton, resorting to a familiar tactic: "Let's enumerate the differences."

Breton: "Both words are used in many different contexts. Let us restrict our consideration to those mathematical contexts which can lead to a physical application.
    Angles can be added numerically. We can add a 90 degree angle to a 30 degree angle to form a 120 degree angle. Directions cannot be simply added to produce another direction."

Einstein, objecting: "Directions are vectors and so can be added in the axiomatic set of vectors."

Breton: "Hold on. Add two angles, obtain another angle. Add two directions, the result is not another direction. Besides, the two additions are different kinds of additions. Let us call the one for angles, scalar addition. The one for directions we can call vector addition. We have not yet defined vector addition, but it must be suitably defined for our vector set."

Einstein thought, 'Breton is objecting with my own objection. I have just used 'addition' ambiguously.' Rather than let Breton underscore that embarrassing point, Einstein quickly took up Newton's agenda: "Angles refer to triangles, whereas directions refer to a unit sphere."

Newton, continuing his agenda: "We can define either in terms of the other.
    An angle can be defined from two directions originating from the same point, called the vertex. The two directions can then serve as sides of a triangle.
    A direction can be defined in terms of angles. First set up a coordinate system of three mutually orthogonal axes. Using the axes as sides, a direction can be defined in terms of three angles."

Breton: "True enough, but consider this, Newton. If Einstein asks me to point at you, I will simply point my index finger in your direction with no reference to angles at all. So directions may be defined in terms of angles, but not necessarily so."

Newton: "Still locations *can* be defined in terms of angles, just as surveyors do. A baseline and two angles are all that is needed."

Einstein, enjoying the different points of view: "But locations are defined even more easily by a direction and a distance."

Breton: "Newton, would a table help us?"

With that Newton happily drew up the following table which he presented to his friends.

Angle	Direction
Scalar addition	Vectorial addition
Refers to triangles	Refers to a unit sphere
Defined in terms of two directions	Does not need angles to be defined, but may be defined in terms of three angles
Can be useful for location	Can be useful for location

Breton, after considering the table,: "The table shows clearly that angle and direction are two different ideas. For a mathematical science like Euclidean Geometry, angle may well be a better choice as prior to direction. The simplicity of direction seems more appropriate for Theoretical Physics."

Turning to his two friends, he questioned:"Do you both agree that angle and direction are ideas, not physical objects?"

Newton: "Of course."

Einstein: "But they may refer to properties of material objects."

Breton: "Location appears a property of all material objects. That is why Theoretical Physics should favor direction as axiomatic in its vector set. So doing, locations are more simply described.

Will you accept then, for our great adventure, that direction is taken as a fundamental and axiomatic idea from which angles may be defined."

Newton quickly agreed, but Einstein commented: "It is a fine and subtle point which I accept reluctantly."

The fineness and the subtlety of the discussion put Breton in mind of a story about proper beginnings.

Peaches

Deliberately, he planted me. He had this vision, one of a prolific peach tree flourishing beside his driveway, just here about five feet off, hiding that view, enhancing this other. He envisioned my spring flowering, the delicately purple blooms before the leaves budded out, a delicately purplish fountain. Then with the coming leaves I would put forth peachlings, little hard nuggets at first, which would grow and grow. With the growing, my branches would begin to bow down, almost to the ground.

He imagined himself sitting in a chair by my trunk, lazily contemplating the peachlings' slow growth. In my shade, he would read, or doze, or simply enjoy a comfortable peace.

The wrens would tell him when to harvest. Brashly, they would pick into the sunny side of the peach, a small indentation, leaving the firmer skin untouched. Time to harvest. A time for calling family and friends. A time for singing, for joyful collecting into baskets, bags, whatever, in which to collect the bountiful harvest. Peaches everywhere, in the kitchen, on the porch, in the fridge, on window sills.

And then the next steps would be launched. Peaches, washed and dried, could be served, whole, drilled or not, or peeled and with stones removed sliced just before serving. Cream could be added as a dessert, or they might find their way into a fruit cup or salad.

Or the peaches could be fried. He would cut the peaches in halves, remove the stones, and cook them over low heat until tender, basting with butter. He might relish the result as a meat accompaniment or even as a desert.

Or they might find their way into delicious peach cobblers, or luscious peach shortcakes or toothsome peach upside down cakes. He could smell the aroma now. Or why not a peach cream pie, or a peach sponge pie. Mouth-watering. Some might be canned, some might be put in jams and jellies, some might be brandied. Visions of peach tarts floated by his imagination.

Thus motivated, he planted me. First he selected me from other seeds in his collection. Then he placed me in a five inch pot filled with potting soil which he watered generously. He smiled when I pushed forth my first leaves. When I grew to six inches, he transferred me from the pot to a large hole in just the location he had in mind. I grew fast. The first year I had grown two feet tall, the next year ten feet tall. He watered, he weeded, he mulched. The leaves, the bark looked exactly like a peach tree. Next year, he smiled to himself, he would be harvesting peaches.

Little does he know, I am an apricot.

Breton: "Perhaps, my dear Einstein, the little story of the apricot will give you a better appreciation of careful beginnings. So let us return to defining vectorial addition."

Newton: "In any case, you want us to accept as an axiom, that
$$\mathbf{v1}+\mathbf{v2} = q(\mathbf{v1}+\mathbf{v2})* \mathbf{u}(\mathbf{v1}+\mathbf{v2}).$$
It seems we can say more about the direction $\mathbf{u}(\mathbf{v1}+\mathbf{v2})$.

Breton: "What do you see?"

Newton: "Let us imagine a plane defined by the two vectors $\mathbf{v1}$ and $\mathbf{v2}$. Suppose further that $\mathbf{v1}+\mathbf{v2}$ lies in the same plane. Then it appears that the direction $\mathbf{u}(\mathbf{v1}+\mathbf{v2})$ equals some ratio of $\mathbf{u}(\mathbf{v1})$ and $\mathbf{u}(\mathbf{v2})$, say
$$\mathbf{u}(\mathbf{v1}+\mathbf{v2}) = a*\mathbf{u}(\mathbf{v1}) + b*\mathbf{u}(\mathbf{v2})$$
for some a and b."

Breton: "And the angle between $\mathbf{v1}$ and $\mathbf{v1}+\mathbf{v2}$ or between $\mathbf{v2}$ and $\mathbf{v1}+\mathbf{v2}$ must always be less that the angle between $\mathbf{v1}$ and $\mathbf{v2}$."

Einstein, always looking to steer the conversation: "Let's do directions as a first step."

Breton: "OK. Suppose two directions, $\mathbf{u1}$ and $\mathbf{u2}$. We know their sum as a vector of the vector set is not a direction. So
$$\mathbf{u1} + \mathbf{u2} = q*\mathbf{u3}$$
Further, $\mathbf{u1}$ and $\mathbf{u2}$ can be thought of a radii of a unit sphere.
So if $\mathbf{u2} = \mathbf{u1}$, what might be an appropriate definition?"

Newton, engaging willingly: "We should have
$$\mathbf{u1} + \mathbf{u1} = 2*\mathbf{u1}.$$

Breton: "And how about $\mathbf{u1} + (-\mathbf{u1})$?"

Newton: "We should have
$$\mathbf{u1} + (-\mathbf{u1}) = \mathbf{0}.$$

Breton: "Now any direction $\mathbf{u2}$ will lie between $\mathbf{u1}$ and $-\mathbf{u1}$, so their corresponding q's will lie between 2 and 0."

Newton: "And their direction?"
Breton: "Half way way between them."

Einstein: "What does that mean?"

Breton: "A few diagrams may be helpful."

With that Breton sketched the following drawings.

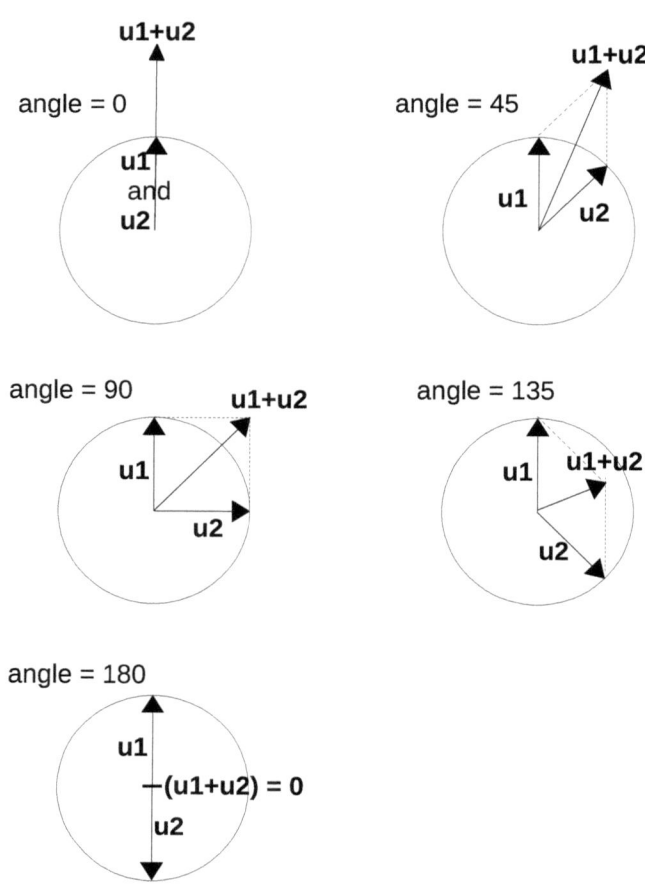

Addition of Directions

Breton: "Look at the drawing carefully. The solid lines with arrows indicate vectors; the dashed lines are parallel to them. In each case a rhombus appears, that is a rectangle (quadrilateral) with four equal sides. The sum of the two directions is indicated by the diagonal of the rhombus. Half of the rhombus formed by **u1**, **u1+u2**, and the parallel **u2**, is a triangle. We can find the length of the sum by the law of projections for triangles."

Einstein: "Show us."

Breton: "The law states
   length(l1) = length(l2)*cos(angle(3))+ length(l3)*cos(angle(2))
where length(l1) is our desired length, length(l2) and length(l3) each equals 1 since they are the lengths of directions. For this case, then the law reduces to
   length(l1) = cos(angle(3))+cos(angle(2))
             = 2*cos(angle(2))
since cos(angle(3))=cos(angle(2)). The angle between the two vectors is just 2*cos(angle(2))."

Newton: "I can put the results into a table."

angle (degrees)	angle(2)	cos(angle(2))	q(u1+u2)
0	0	1	2
45	22.5	0.92388	1.846776
90	45	sqrt(2)/2	1.41422
135	67.5	0.38638	0.77276
180	90	0	0

Einstein: "Justify your formula for q(**u1+u2**)!"

Breton: "This is simply an exercise in trigonometry. Follow along in this diagram."

With that Breton handed his two friends the following diagram.

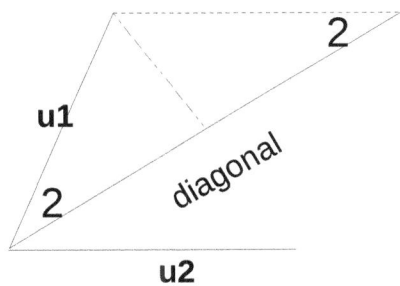

Breton: "First from the tip of the first direction, drop a line perpendicular to the diagonal line, as shown. Then note that the perpendicular line divides the larger triangle into two other equal right triangles. Moreover, the angle between the first direction and the diagonal is just half the angle between the vectors. Now in such a configuration the cosine of an angle is defined as the ratio of the hypotenuse to the adjacent side, that is
          cos(angle/2) = (diagonal/2)/1.

Therefore,
$$\text{diagonal} = 2 * \cos(\text{angle}/2)$$
$$= 2 * \cos(\text{angle}(2)).$$
as I have indicated above."

Einstein, returning to the attack: "It appears to me that by using triangles and perpendiculars you have embraced Euclidean geometry."

Newton, unwilling to neglect such an opportunity: "And given a good reason for the instinctive genius of my illustrious forebear in basing his Physics on Euclidean geometry."

Breton, retreating: "Einstein is right. By insisting on measuring the diagonal, I have lost the path. Theoretical Physics should not be tied to Euclid's geometry, or indeed to any geometry at all. Nor should our vector set. I have made specific what might well have been left unspecified. Still the process of imagining the sum of two vectors from the image of a rhomboid can stand, provided we do not tie the rhombus to a Euclidean plane."

Newton: "You are denigrating my illustrious ancestor."

Breton: "Not only yours, but Einstein's as well."

Einstein, still prodding: "Some sort of geometry *has* to be assumed for Physics."

Breton: "And if it doesn't correspond with reality?"

Unable to respond both Newton and Einstein fell silent.

Breton: "We are engaged in an effort to create ideas which correspond to physical reality. To geometrize the description of location may impose an assumption which leads physicists astray. Theoretical Physics needs only to conceive of a vector set with vectorial addition satisfying the axioms. Addition in the vectorial set is merely *illustrated* by the diagonal of the rhombus."

Einstein: "I find it difficult to think about location without a geometry."

Breton: "Just as you found it hard to imagine immaterial ideas. Still, to describe location as a length and a direction requires no assumption of a specific geometry."

Newton: "But its measurement will."

Breton: "Very likely, but measurement is of little concern to the science of Physics. See how easily we slip into technology and away from science."

Newton: "Why vectorial addition in the first place?"

Breton: "We observe physical objects as mutable. An object extended in one location may subsequently be observed in a different location; an object moving in a certain direction, may subsequently be observed moving in another direction. An object being forced in one direction may subsequently be forced in another direction. A mathematical vector set has the possibility of being transformed into an appropriate concept for Theoretical Physics to describe and understand these observations."

Einstein, expansively: "So my illustrious ancestor conceived the idea of space-time as an idea with which to describe the interactions space and time in the universe."

Newton: "Ridiculous! Breton does our vector space validate space-time vectors?"

Breton, accepting the challenge: "Tell me: Is space something physical, or is it an idea?"

Einstein: "Physical!"

Newton: "Physical!"

Breton, moving the argument: "Is a triangle in Euclidean geometry physical or ideal?"

Newton: "Ideal, but it can be applied to physical objects."

Einstein, obstinately: "A triangle is extended, just as space is extended."

Breton, continuing: "As an idea, is the idea of a triangle itself extended or is it an idea of an extended mathematical object?"

Einstein, somewhat intrigued: "How can an idea be physically extensible?"

Breton: "Remember yesterday in tp1.1 the discussion of human cognition—how we can distinguish between a physical object, its symbol, and the idea of the object? An illustration of a triangle symbolizes the idea of a triangle. If we mean the illustration, then we see the illustration as an extended figure. If we mean the idea, then the idea itself is not extended."

Newton, objecting: "Triangles can be said of physical objects!"

Breton: "Properly or improperly? We can define a physical object as a triangle, but then the physical triangle differs from the geometrical triangle. We are then using the word 'triangle' ambiguously. Or we can describe the physical object as having some attributes of a geometrical triangle. Then we should say more accurately the physical object is like a triangle, that it is triangular."

Einstein, returning to the main subject: "So there is physical space, symbols for space, and the idea of space. What is the symbol for space?"

Breton: "Or is the symbol for space confused with physical space? What is physical space?"

Einstein, unwilling to answer: "You tell me."

Breton: "We have agreed that the science of Physics deals with objects that are extended, moving or forcing. Physical space would have to be an object which is extended, moving or forcing."

Einstein, with an uncertain note of triumph: "Space is extended."

Breton: "But nothing more. Would you describe it as an extended nothingness."

Einstein: "Only reluctantly."

Breton, explaining the obvious: "It may be extended, but it is not an existing physical object. It is rather an idea which can be useful in describing material objects and their positions."

Einstein, exclaiming: "You are undermining accepted insights developed by my illustrious ancestor!"

Newton, quickly siding with Einstein: "Not only yours, but mine as well."

Breton, soothingly: "Neither of your illustrious ancestors had the advantages of the well developed language of Theoretical Physics. Had they been so fortunate, they may not have fallen into the error of reifying space or time. I know this conclusion must shatter your images of your illustrious ancestors, both great men, both subject to error. You may appreciate the following conversation."

SPACE-TIME

Professor says:
For describing the universe, nothing serves so well as vectors. What is a vector? A vector consists of row of numbers in given order. Each of the positions in the order is called a coordinate, also called a dimension. Each of the dimensions houses a real number. For instance, to describe positions in space we use three dimensions.

Student says:
Since the universe can be described by three infinite sets of numbers, it too must be infinite.

Professor says:
That's the beauty of vectors; we can always add another dimension. In the case of the universe we add a fourth dimension called time. This dimension can react with the spatial dimensions. At one time the universe may be described by a very small point. At another it could be described as something much larger. At present the spatial dimensions can be restricted to the present universe.

Student says:
Why can't the spatial dimensions react with each other? Suppose two of the spatial dimensions contain only one fixed number, Then the description of the universe becomes one straight line, no matter what the time dimension contains.

Professor says:
That's just the way it is. The spatial dimensions are independent of each other, but dependent on the temporal dimension.

Student thinks:
Why not add a further dimension and call it the NONSENSE dimension.

Einstein, unwilling to surrender: "We will have to take up this matter again. For now, let us return to vectorial algebra."

Newton: "We were discussing vectorial algebra, and had described addition of two directions."

Einstein: "How do you finally describe vectorial addition for *any* two vectors in our special mathematical vector set?"

Breton: "First let me review what we have learned from directions. The specific definition of addition for *directions* fits some of the axioms of a vector set. Let me list them.
    u1 + u2 = u2 + u1
    1∗(**u1** + **u2**) = 1∗**u1** + 1∗**u2**
    (1+1)∗**u1** = 1∗**u1** + 1∗**u1**
    1∗**u** = **u**, for any direction
and for every direction **u**, there exists a vector −**u** such that
    u + (−**u**) = **0**

Einstein: "But not all the axioms are satisfied."

Breton: "True enough. Remember we started the investigation of the plus operator for *vectors* by first considering what might be appropriate for *directions*. Now we can climb a little higher to consider addition for vectors generally."

Newton: "I propose a simple extension of our results for directions. We can take
        v1+v2 ≡ q1∗**uv1** + q2∗**uv2**
from a *rhomboid* rather than a *rhombus*."

Einstein, pugnaciously: "Don't go hiding behind some fancy names. Explain each and show us how they differ."

Newton: "A rhombus is a quadrilateral with four equal sides. A rhomboid is a quadrilateral two of its sides not necessarily equal in length but matched by equal, parallel sides. An illustration can bring out the difference perhaps more clearly than words. Breton, would you kindly draw us a rhombus and a rhomboid."

Breton quickly obliged with the following drawings.

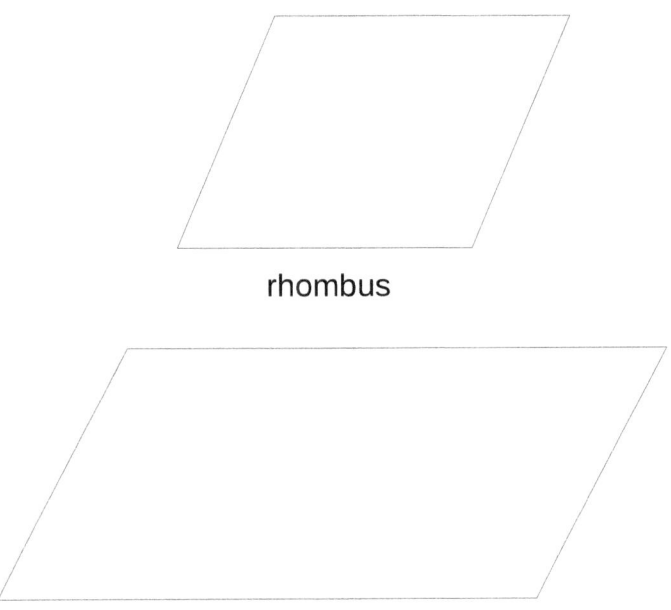

rhombus

rhomboid

Newton: "For directions we used a rhombus each of whose sides had a length equal to one. Then the addition of two vectors was defined as the diagonal of the rhombus.
 We can extend that definition to any two vectors of equal length.
$$q \bullet u1 + q \bullet u2$$
by referencing a *rhombus* the length of whose sides equals q.

We can finally extend the definition to any two vectors $q1 \bullet u1 + q2 \bullet u2$ by referencing a *rhomboid* the length of whose sides equals q1 and q2."

Breton: "Then vector addition can be referred in all instances by the diagonal of a rhomboid. Here is a diagram which illustrates vectorial addition generally."

With that Breton handed his friends the following illustration.

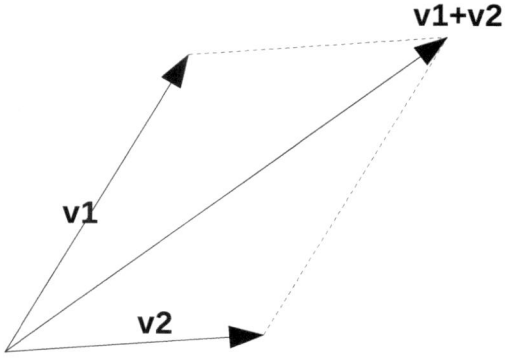

vector addition

Einstein: "The drawing shows what you are trying to define, but what is the length of **v1+v2**?"

Breton: "Here this sketch may help you."

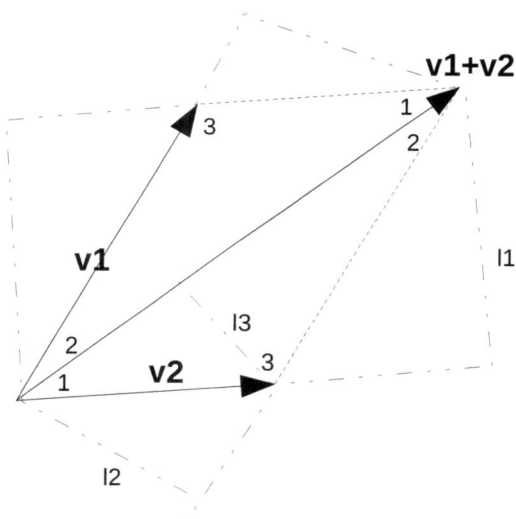

Breton: "This sketch labels the three angles: 1, 2, and 3, and includes extension lines so the their sines can be indicated. Angle1 lies opposite **v1**; angle2 lies opposite **v2**; angle3 lies opposite **v1+v2**. Since
sin(angle1) = length(l1)/length(**v1+v2**),
    length(**v1+v2**) = length(l1)/sin(angle1).

Since
sin(angle2) = length(l2)/length(**v1+v2**),
   length(**v1+v2**) = length(l2)/sin(angle2).
Both angle1 and angle2 are acute angles, but angle3 is obtuse. Referring to angle3, the length of the sum can be expressed in terms of cosines.
   length(**v1+v2**) = cos(angle1)/length(**v2**)
            + cos(angle2)/length(**v1**)
So here Einstein are three equations for length(**v1+v2**)."

Newton: "Triangles seem to be implied in this definition of vectorial addition. Can we say more about relationships with triangles."

Breton: "We might say a great deal more, which would carry us towards geometry rather than vectorial algebra. But to satisfy your love of geometry, Newton, let me prove a relationship between sides and angles in a triangle. The following relationship, called the **sine law of oblique triangles** may prove useful in the future
. *In any oblique triangle, the sines of each angle are proportional to their opposite sides.*"

Einstein: "Show us a diagram."

With that request, Breton quickly produced the following sketch.

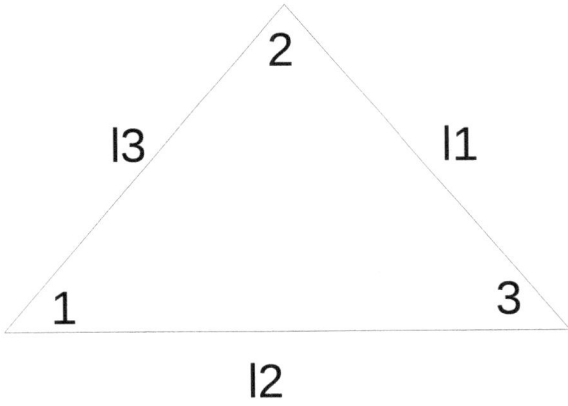

Breton: "The angles are labeled 1, 2, and 3. As you can see the sides opposite the angles are marked l1, l2, and l3 respectively."

Einstein: "That looks just fine in the nicely proportioned triangle you have drawn. But you claim the relationship holds for any oblique triangle."

Breton: "The proof for the law of sines will not rely on any two angles or sides begin equal."

Newton: "Then prove the sine law."

Breton: "The law of sines states that
(sin(angle1)/length(l1) = (sin(angle2)/length(l2)
                        = (sin(angle3)/length(l3).

Newton: "Where are sines in your illustration?"

Breton: "To see the sines, construct orthogonal lines from the apex of each angle to its opposite side as illustrated in this diagram.

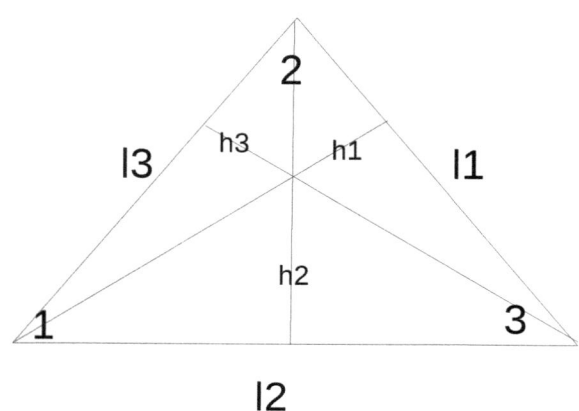

I have labeled the orthogonal lines as h1, h2, and h3.
Then the sine of each of the angles has two representations.

sin(angle1) = length(h3)/(length(l2) = length(h2)/(length(l3)
sin(angle2) = length(h1)/(length(l3) = length(h3)/(length(l1)
sin(angle3) = length(h2)/(length(l1) = length(h1)/(length(l2)

Notice that sin(angle2) has a representation containing (length(l1). Likewise sin(angle3). So we can use these equations to form the following ratios.
(sin(angle1)/length(l1)
    = (length(h3)/(length(l2))*(sin(angle2)/(length(h3))
    = (sin(angle2)/(length(l2))
and
(sin(angle1)/length(l1)
    = (length(2)/(length(l3))*(sin(angle3)/(length(h2))
    = (sin(angle3)/(length(l3))
And this my dear Einstein, proves the law of sines for any oblique triangle."

Einstein, nodding agreement quickly redirects the conversation to vectorial algebra: "Express the difference between two vectors!"

Breton: "I will have to expand my former sketch a little."

With that Breton quickly handed his friends the following sketch.

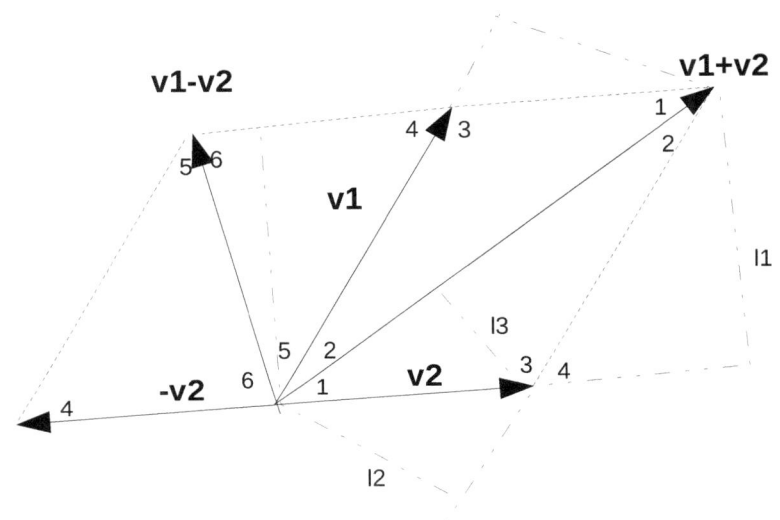

Breton: "Now three other angles have been indicated: angle4, angle5, and angle 6. From the sketch we see
  sin(angle4)/length(**v1−v2**) = sin(angle5)/length(**−v2**)
                                = sin(angle6)/length(**v1**)
so that
length(**v1−v2**) = length(**−v2**)∗sin(angle4)/sin(angle5)
length(**v1−v2**) = length(**v1**)∗sin(angle4)/sin(angle6)
   The answer may involve us again in a specific geometry and lead us off our chosen path. For our purposes we will simply accept as axiomatic that our vector set has an addition operator which operates on any two vectors as referenced in our rhomboid illustration without implying a specific geometry."

Einstein: "So what can you give for a *definition* of vectorial addition."

Breton: "Nothing. Addition in the vector set is an axiomatic assumption. It can be described, but not defined since definition would imply something 'more' axiomatic."

Newton, cooperatively: "In Euclidean Geometry, a 'line' is an axiomatic assumption. It cannot be defined in terms of simpler axioms, but merely accepted and described."

Breton: "So by the axioms we are given a plus operator which operates on any two vectors in the vector set as
             +(**v1**,**v2**) = **v1**+**v2**.

Newton: "Specifically, if **v2** = **0**,
$$v1 + 0 = v1.$$

Einstein, agnosticly: "Show me."

Newton: "Look at the diagram. As **v2** goes to **0**, angle 2 also becomes zero and **v1+v2** becomes **v1**."

Einstein: "And if **v2** = **−v1**?"

Newton: "Look at the diagram again. Let **v2** become **−v1**. Then angle 2 plus angle 1 equals pi and **v1+v2** becomes **0** orthogonal to **v1**. ...So we see the rhomboid scheme leads to a description of the **+** vectorial operator consistent with the axioms for the vectorial set.
  The plus operator acts symbolically like the plus operator for integers."

Breton: "So let us accept that a set symbolized as
  V ={{q∗**u** such that q is an element of Q, **u** a direction},+}
as a vector set. since it satisfies all the axioms of a mathematical vector set."

Newton: "I agree."

Einstein: "I also agree, but where is all this leading to?"

Breton: "Remember how we developed the quotient numbers, starting with the positive integers? We moved from the positive integers, to the negative integers, to multiplication, to division, each time enlarging our consideration to a set which finally supported an algebra. Then we showed that the set of quotient partitions could support a topology from which we could define limits and then continuous functions. Do you think our vector set could support a similar development?"

Einstein: "You put before us an ambitious agenda. Who knows where it will end?"

Newton: "If we follow our earlier development we might expect surprises."

Breton: "And adventure, intellectual adventure."

Subtraction in the axiomatic set of vectors

Newton: "Subtraction is easy. From its axioms the vector set already contains a vector **−v** for every vector **v**. Moreover, as we have seen in the illustration for addition in the vector set, we accept that
$$v + (-v) = 0.$$

Breton: "Can minus act like an operator?"

Newton: "The axioms give us *plus* as an operator, but not *minus*. If minus were an operator we would need to know
$$v1 - v2$$
for any **v1** and **v2**."

Breton: "We already know
$$0 + (-v2) = -v2$$
which we could take as
$$0 - v2 = -v2.$$

Newton: "Minus would be well defined as an operator as
$$v1 - v2 \equiv v1 + (-v2).$$

Einstein: "Breton, show us how this would look as an illustration."

Breton: "Gladly."

With that he quickly produced the following drawing.

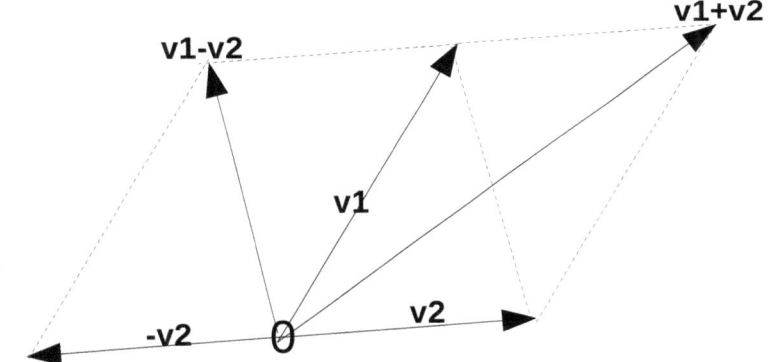

Vector Addition and Subtraction

Breton: "Using the same rules of vectorial addition, you can see that **v1−v2** may differ from **v1+v2** in both length and direction."

Newton: "Notice the line parallel to **v1−v2**, the one stretching from the tip of **v2** to the tip of **v1**. It has the same length and direction as **v1−v2**."

Einstein: "Same length, but not a direction since it does not relate to the unit sphere."

Breton: "It would if the unit sphere were centered at the tip of **v2** instead of **0**."

Newton: "Look at the dashed line parallel to **v2**. We would arrive at **v1+v2** by traveling along **v1** and the parallel line."

Breton: "And we would arrive at **v1−v2** by traveling along **v1** and the line parallel to **−v2**. Have you discovered a new way for defining addition and subtraction in the vector set?"

Newton, proudly: "Yes we have. I am tracing some paths in the diagram. They all comply with the rule.

Breton: "So the parallel lines can be thought of as translated base vectors. Allowing translated vectors enables vectors to be added and subtracted. For instance,
$$v1+v2 = v1 + v2.$$

Einstein: "Not always. The direction of the translated vector seems important. For instance take the path
$$v1+v2 + v1$$
does not equal **v2**.

Breton: "You're right, but look
$$v1+v2 - v1 = v2$$
works fine. So taking the path in the direction opposite to the base vector produces a vectorial subtraction. So we can define sums of vectors analogously to sums of numbers. The start of each of the summands will be the tip of the arrow of its previous member. The summand will be plus or minus depending on its correspondence to the direction of its base vector."

Newton: "More than analogous. If we take the partitions of Q as a vector set with only two directions, plus and minus, then the analogy becomes perfect. So we have achieved a generalization of numerical Arithmetic for our vector set."

Breton: "Another of your splendid insights. But our notation has not followed this new way of vectorial Arithmetic. Let me propose a similar expansion of our notation. Our vectors have been designated as
$$\mathbf{v} = q(\mathbf{v})*u(\mathbf{v}).$$
Each translated vector can be written as
$$\mathbf{v} = \mathbf{v+v0} - \mathbf{v0}$$
one for each **v0**. That is,
$$q(\mathbf{v})*u(\mathbf{v}) = q(\mathbf{v+v0})*u(\mathbf{v+v0}) - q(\mathbf{v0})*u(\mathbf{v0})$$
So for a translated vector we use
$$\mathbf{v} = \mathbf{v2} - \mathbf{v0}$$
where **v2** = **v+v0**. A translated vector of **v** can be thought of as starting from a base vector **v0** and extending in the direction $u(\mathbf{v})$ for a length $q(\mathbf{v})$ to the vector **v2**. A translated vector can be found for each base vector **v0**."

Einstein: "Then our former notation can be seen as having implied **v0** = **0** for the base vector."

Newton: "Something like the partitions of Q. A given vector and all its translations acts like a partition in the vector set, one for each value of q and each direction."

Breton: "Whereas direction was muted concept in Q, it becomes prominent in our vector set."

Einstein: "It's uncanny. The partition containing 1/1 in Q differs from the partition containing −1/1, just as the partition in **V** for 1∗**u**(**v**) differs from the partition for −1∗**u**(**v**)."

Newton: "Even more. The partition containing −1/(−1) contains 1/1, just as the partition in **V** for −1∗**u**(−**v**) contains the vector 1∗**u**(**v**)."

Breton: "So we have achieved an intellectually beautiful vista. Though different, the partitions of Q reveal a similarity to the partitions of **V**. The perception of an underlying unity gives us a better appreciation of both and brings us intellectual pleasure."

The three friends fell silent contemplating this revelation of unity in complexity. Finally, Breton broke the silence.

Breton: "Unity in complexity. This trait seems a most desirable quality of any science. It seems to me, however, the trait applies to a much larger reality than a science like Physics. By your leave, let me reflect on this a little.
 We have proven God's existence. Beyond his existence, we acknowledge that our knowledge of God would be greatly augmented were he to reveal himself."

Newton, intrigued by Breton's remark: "Of course, but many false claims could also arise."

Breton: "Most certainly. Given that a revelation has truly occurred, we can be assured that many false imitations would arise, many corruptions, many masqueraders to obscure the true revelation. Just as with the science of Physics, many false claimants will arise. It is the human condition. Nevertheless true sciences exist, such as arithmetic and geometry. A true science of Physics is indeed possible, which is the reason for our efforts with Theoretical Physics. In a similar way, false revelations do not diminish the importance of the true revelation. Moreover, the true revelation should show signs of its veracity.
 For our reflection let us consider the Ark of the Old Testament. The story is told in the book of Exodus.
 God, in his goodness, chose to reveal himself to the descendants of Israel through Moses to whom He would give two tablets of stone on which God would write ten commandments. Consequently they would be sacred objects. Treating them casually or disrespectfully would risk God's wrath. Now Moses had been meeting God in a tent set apart, where he and only he would enter. When so engaged a cloud would cover the Tent of Meeting.

 During these conversations, God commanded Moses in detail to build a chest (ark) to house the tablets.
 The Lord commanded Moses to frame an ark of setim wood. The

wood grows as acacia trees in the desert wadis, or torrent valleys, of Sinai. The wood is light, hard, and durable, and grows almost as black as ebony with age. The Ark was to be 2½ cubits in length, 1½ cubits in breadth, and 1½ cubits in height (approximately 131×79×79 cm or 52×31×31in). Then it was to be gilded entirely with gold, and a crown or molding of gold was to be put around it. Four rings of gold were to be attached to its four corners, two on each side—and through these rings staves of setim-wood overlaid with gold for carrying the Ark were to be inserted; and these were not to be removed.

Moses was commanded to cut two tablets of stone and place them in the ark. God would inscribe the ten commandments on them. Moses was also to make a propitiatory of the purest gold with two cherubim of the purest gold, one cherub to be placed on one side of the Ark, the other on the other side. The cherubim were to cover both sides of the propitiatory spreading their wings and covering the oracle. The Ark finally was to be placed under a veil.

God promised to speak to Moses over the propitiatory and from the midst of the two cherubim

The meeting tent had been placed apart from the encampment while the ark and the other furnishings of the tent were being designed and constructed. But on the day the tabernacle was completed the Israelites marched their portable cathedral to the exact center of the camp with their tents and personal belongings radiating from the Ark of the Covenant. The tabernacle gave the Israelites a visible reminder of God's central place. Each day priests performed functions of sacrifice and worship there. The portable cathedral was God's house on Earth. The sanctuary of the cathedral the Holy of Holies was the most sacred place in the universe. At its very center was the ark of the Covenant that mysterious golden chest covered by a propitiatory with two cherubim.

The biblical account continues that, after its creation by Moses, the Ark was carried by the Israelites during their 40 years of wandering in the desert.

In the desert the cloud (the Shekinah) overshadowed the tabernacle of the testimony and the glory of the Lord filled it. And when the priests came out of the sanctuary a cloud filled the house of the Lord. This overshadowing indicated the very presence of God both in the ark and in the Tent.

With the Ark God entered the human condition and so it acquired a history.

When the Israelites, led by Joshua toward the Promised Land, arrived at the banks of the Jordan river, the Ark was carried in the lead preceding the people and was the signal for their advance. During the crossing, the river grew dry as soon as the feet of the priests carrying the Ark touched its waters, and remained so until the priests—with the Ark—left the river after the people had passed over.

After the defeat at Ai, Joshua lamented before the Ark.

When Joshua read the Law to the people between Mount Gerizim and Mount Ebal, he did so in the presence of the Ark.

We next hear of the Ark in Bethel where it was being cared for by the priest Phineas, the grandson of Aaron.

Later the Ark was kept at Shiloh at the time of the prophet Samuel's apprenticeship, where it was cared for by Hophni and Phinehas, two sons of Eli.

According to the Biblical narrative, a few years later the elders of Israel decided to take the Ark out onto the battlefield to assist them against the Philistines. They were heavily defeated with the loss of 30,000 men. The Ark was captured by the Philistines and Hophni and Phinehas were killed.

The Philistines took the Ark to several places in their country, and at each place misfortune befell them. At Ashdod it was placed in the temple of Dagon. The next morning Dagon was found prostrate, bowed down, before it; and on being restored to his place, he was on the following morning again found prostrate and broken. The people of Ashdod were smitten with tumors; a plague of mice was sent over the land. The affliction of boils was also visited upon the people of Gath and of Ekron, whither the Ark was successively removed. After the Ark had been among them for seven months, the Philistines, on the advice of their diviners, returned it to the Israelites, accompanying its return with an offering consisting of golden images of the tumors and mice wherewith they had been afflicted. The Ark was set up in the field of Joshua the Beth-shemite, and the Beth-shemites offered sacrifices and burnt offerings. Out of curiosity the men of Beth-shemesh gazed at the Ark; and as a punishment, seventy of them were smitten by the Lord..

The Bethshemites sent to Kirjath-jearim to have the Ark removed; and it was taken to the house of Abinadab, whose son Eleazar was sanctified to keep it. Kirjath-jearim remained the abode of the Ark for twenty years. Under Saul, the Ark was with the army before he first met the Philistines, but the king was too impatient to consult it before engaging in battle.

In the Biblical narrative, at the beginning of his reign over the United Monarchy, King David removed the Ark from Kirjath-jearim amid great rejoicing. On the way to Zion, Uzzah, one of the drivers of the cart that carried the Ark, put out his hand to steady the Ark, and was struck dead by God for touching it. David, in fear, carried the Ark aside into the house of Obed-edom the Gittite, instead of carrying it on to Zion, and it stayed there for three months.

On hearing that God had blessed Obed-edom because of the presence of the Ark in his house, David had the Ark brought to Zion by the Levites, while he himself, "girded with a linen ephod danced before the Lord with all his might" and in the sight of all the public gathered in Jerusalem. In Zion, David put the Ark in the tent he had prepared for it, offered sacrifices, distributed food, and blessed the people and his own household. David used the tent as a personal place of prayer.

The Ark was with the army during the siege of Rabbah; and when David fled from Jerusalem at the time of Absalom's conspiracy, the Ark was carried along with him until he ordered Zadok the priest to return it to Jerusalem.

According to the Biblical narrative, when Abiathar was dismissed from the priesthood by King Solomon for having taken part in Adonijah's conspiracy against David, his life was spared because he had formerly borne the Ark. Solomon worshiped before the Ark after his dream in which God promised him wisdom.

During the construction of Solomon's Temple, a special inner room, the Holy of Holies, was prepared to receive and house the Ark; and when the Temple was dedicated, the Ark—containing the original tablets of the Ten Commandments—was placed therein. When the priests emerged from the holy place after placing the Ark there, the Temple was filled with a cloud, "for the glory of the Lord had filled the house of the Lord".

When Solomon married Pharaoh's daughter, he caused her to dwell in a house outside Zion, as Zion was consecrated because it contained the Ark.

King Josiah also had the Ark returned to the Temple, from which it appears to have been removed by one of his predecessors.

In 587 BC, the Babylonians destroyed Jerusalem and Solomon's Temple. There is no record of what became of the Ark in the Books of Kings and Chronicles.

What happened with the old ark is reported in 2 Maccabees chapter 2. Jeremiah, warned by God, commanded that the tabernacle and the ark should accompany him to the mountain, the mountain where Moses went up and saw the inheritance of God. The mountain that is spoken of is Mount Nebo.

The Ark of the Old Covenant is holy. It is the sign of God's presence among His people. God overshadows the Ark and as a consequence the Ark carried within it the word of God in stone. Those who receive the Ark are blessed.

Consider the Blessed Virgin Mary, mother of Jesus the Christ, as the Ark of the New Covenant. Whereas the old Ark was constructed of inanimate materials, the Ark of the New Covenant is Mary's sacred womb in which the Word of God is made flesh. The Ark of the Old Covenant carried the word of God in stone; the Ark of the New Covenant carried the Word of God made human. The divine maternity is implicit in the Ark of the Old Covenant. Consider the beautiful detail in the overshadowing to describe the way in which the Son of the Most High would be incarnate. 'The Holy Spirit shall come upon you and the power of the Most High shall overshadow you.'

Just as the old Ark was mobile, so too is the New Ark. At the suggestion of Angel Gabriel, Mary visits with haste her pregnant cousin Elizabeth where she is greeted as mother of the Lord. Her presence becomes a blessing and a cause of great joy. From henceforth all generations will call her blessed. She becomes the sign of God's true presence for the sanctification of His people. She is blessed by all generations of God's children. The early Church referred to her as the Theotokos, the God bearing one.

When Our Lady brought the newborn Jesus to be presented at the temple, the old Ark was not there. Mary, the Ark of the New Covenant, had come into the Temple.

Again and again, manifold events in Mary's life find a prefigurement in the Ark of the Old Testament and, amazingly, events of a millennium previous are realized again surprisingly in Mary's life. The surprising concurrence over time lends credibility that the narrative is not a false claim.

What the ark meant in symbol, Mary is in reality. But the ark was only a symbol; in Mary, God was really present so much so as to become her Son.
 Finally at the end of her life Mary was assumed bodily into heaven, much as the old Ark has been removed from human view."

Newton: "You have wandered far from Theoretical Physics."

Breton: "In a way perhaps, but not so far as you might imagine.. The science of Physics, recall, is reductive, that is, it deals with a small aspect of reality, not the whole. When I talk of Mary, mother of Jesus Christ, I refer to a larger reality which comprehends and gives meaning to Physics. The larger reality acts like a frame of, or as a background to, physical reality. It is an unfortunate error to expect Physics to comprehend the larger reality. Its like defining a function in the wrong direction."

Newton: "Don't you agree that the study of Physics, or even of Theoretical Physics, differs from the study of Mary?"

Breton: "Of course. Still Mary is the mother of Jesus for whom all of reality has been created, including physical reality. She can thus be honored as queen of the universe, and deserves a place in our thoughts. I for one, take Mary as patron of our efforts to explore the origins of Physics. I like to invoke her help especially when the trail becomes difficult."

Einstein: "Invoke as you will, but let us continue our quest for a vectorial algebra. Can we multiply in the vector set?"

Multiplication in the axiomatic set of vectors

Breton: "Multiplication is not included in the axioms."

Newton: "Then let us define it."

Breton: "Unlike the addition of vectors which produces another vector, multiplication according to our rules for physical units can produce objects which are not vectors."

Einstein: "Why be restricted to rules for labels when we are defining a mathematical structure?"

Breton: "We are looking to define mathematical objects which can be transformed into Theoretical Physics. So it makes sense to respect the restrictions even in mathematics. Remember how we just used the same principle when we refused to add scalar variables with vector variables."

Newton, impatiently: "Agreed. So how do we proceed?"

Breton, plodding forward: "Let's enumerate the possibilities. The product resulting from the multiplication of two vectors could be a member of the underlying field, Q. Or again, it could be another vector."

Einstein: "Then these would be two different kinds of multiplication."

Breton: "Correct. And let me add still another product, a transformation."

Newton: "What kind of transformation?"

Breton: "The transformation would take one vector and transform it into another vector. Although involving vectors, the transformation itself is not a vector."

Einstein: "So far I hear words; show us concretely what you mean?"

Breton, patiently: "Will this help? Multiplication is a kind of function. So if I describe the domain and ranges, you may perceive what I am saying more clearly:
$$\text{multiplication1}: \mathbf{V x V} \rightarrow \mathbf{Q}$$
$$\text{multiplication2}: \mathbf{V x V} \rightarrow \mathbf{V}$$
$$\text{multiplication3}: \mathbf{V x T} \rightarrow \mathbf{V}.$$

Einstein: "So the vector set you propose has four different kinds of multiplications: scalar multiplication which we accept axiomatically, and then these other multiplications. I can see that the additional multiplications are each different because they have different ranges. But they are still undefined."

Inner (Dot) Product

Breton: "So let us start with multiplication1.

Definition (inner (dot) product)
Given
        v1, **v2** vectors in the axiomatic set of vectors.
for
        v1 = q(**v1**)∗**u**(**v1**)
        v2 = q(**v2**)∗**u**(**v2**)
        angle(1,2), the angle between **u**(**v1**) and **u**(**v2**)
then
        v1 • **v2** ≡ q(**v1**)∗q(**v2**)∗cos(angle(1,2))

                                        end of definition

Multiplication1 is called the **inner product** or sometimes the **dot product**. By this curious convention, we call the function, •, by its image. To avoid confusion with the other multiplications in the vector set, it is symbolized with '•'.
As you can see
$$\bullet : \mathbf{V} \mathbf{x} \mathbf{V} \to Q$$
where **VxV** is the joint set."

Einstein, analytically: "The product depends on the angle between the two vectors."

Breton: "Correct. Suppose both vectors are unit vectors. What would be the result?"

Einstein: "Then
$$\mathbf{u}(\mathbf{v1}) \bullet \mathbf{u}(\mathbf{v2}) = 1 * 1 * \cos(\text{angle}(1,2))$$
so cos(angle) equals the inner product of the directions of the two vectors. From this we can conclude that
$$\mathbf{u}(\mathbf{v1}) \bullet \mathbf{u}(\mathbf{v2}) = \mathbf{u}(\mathbf{v2}) \bullet \mathbf{u}(\mathbf{v1})$$
and even
$$\mathbf{v1} \bullet \mathbf{v2} = \mathbf{v2} \bullet \mathbf{v1}.$$

Breton: "What would be the result if the angle were 0?"

Einstein: "Since cos(0) = 1,
$$\mathbf{v1} \bullet \mathbf{v2} = q(\mathbf{v1}) * q(\mathbf{v2}).$$

Breton: "And if **v1** = **v2**?"

Einstein: "Then
$$\mathbf{v1} \bullet \mathbf{v1} = q(\mathbf{v1}) * q(\mathbf{v1}).$$

Breton: "So then the inner product of a vector with itself is equal to the square of its length."

Einstein: "Interesting."

Breton: "Suppose angle equals 90 degrees."

Einstein: "Then cos(angle) = 0, so
$$\mathbf{v1} \bullet \mathbf{v2} = 0.$$

Breton: "Two vectors so related are said to be **perpendicular** to each other, also called **orthogonal** vectors.
  And if the angle equals 180 degrees?"

Einstein: "Then cos(angle) = –1, so
$$\mathbf{v1} \bullet (-\mathbf{v2}) = -q(\mathbf{v1}) * q(\mathbf{v1}).$$

Breton: "Which would be the same as
$$(-\mathbf{v1}) \bullet \mathbf{v2} = -q(\mathbf{v1}) * q(\mathbf{v2}).$$

Einstein: "Correct."

Breton: "So the inner product varies from $q(\mathbf{v1}) * q(\mathbf{v2})$ to $-q(\mathbf{v1}) * q(\mathbf{v2})$ depending on the alignment of the two vectors."

Einstein: "This inner product can be a very interesting addition to our vector set."

Breton: "Yes indeed. The enhancement will become even more interesting when we discuss how to transform it into Theoretical Physics.
  We can come to appreciate the inner product more by considering its geometrical rendition. Obviously from its definition
$$\mathbf{v1} \bullet \mathbf{v2} = \mathbf{v2} \bullet \mathbf{v1}$$
since both equal $q(\mathbf{v1}) * q(\mathbf{v2}) * \cos(\text{angle}(1,2))$.
  Now look how this plays out geometrically."

With that Breton handed the following sketch to his friends.

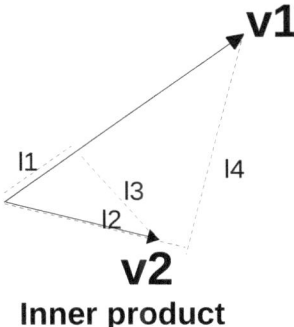

Inner product

Breton: "Along with the two vectors, the sketch shows two right triangles composed of the lines: l2, l4, q(**v1**) and l1, l3, q(**v2**). From the sketch
$$\cos(\text{angle}(1,2)) = \text{length}(l2)/q(\mathbf{v1}) = \text{length}(l1)/q(\mathbf{v2}).$$
Now consider
v1 • **v2** = q(**v1**)*q(**v2**)*cos(angle(1,2))
    = q(**v1**)*q(**v2**)*length(l2)/q(**v1**)
Similarly
v1 • **v2** = q(**v1**)*q(**v2**)*length(l1)/q(**v2**)
So
$$q(\mathbf{v2})*\text{length}(l2) = q(\mathbf{v1})*\text{length}(l1)$$
a result somewhat difficult to visualize geometrically. Thus we can often use a result easily proved vectorially to establish a result much more difficult to prove geometrically. And vice-versa."

Newton, again cooperatively: "Since both q(**v**)'s and l's are lengths, when we talk about their products we are talking about areas. The areas are different, but they have the same value. To prove their equivalence geometrically we would have to slice up one area into pieces which could be superimposed on the second area."

Einstein, hoping to cut the discussion short: "It would be easier to measure both."

Breton, countering: "But the measurement would always be inexact, so by measurement we could never prove the areas were exactly equal."

Newton: "And we would have had to choose some unit of measurement."

Einstein: "Breton, show us a sketch of the area of an inner product."

Complying, Breton produced the following sketch.

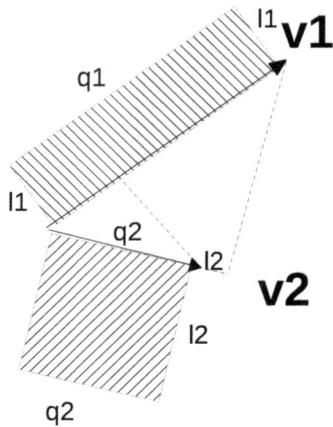

Breton: "Each hatched areas equal **v1** • **v2**."

Breton, with a note of urgency: "So we come to appreciate the beauty and harmony of the inner product. But let's move on."

After a short pause Breton continued in an agreeable tone. "I suggest we simplify our notation. Let us write
- q1 for q(**v1**)
- q2 for q(**v2**)
- q3 for q(**v3**)
- **uv1** for **u(v1)**
- **uv2** for **u(v2)**
- **uv3** for **u(v3)**

Whenever no ambiguity will follow, we can do the same in other contexts."

Einstein, joining gladly: "Agreed."

Newton, with a note of delightful satisfaction: "We are building a mathematical structure--the parts fit together."

Breton, equally satisfied: "We've become intellectual carpenters."

Cross product

Newton: "How about multiplication2?"

Breton: "Again we need a definition." So Breton quickly produced the following definition for his friends.

Definition (cross product—initial)
 Given
  v1, v2 vectors in the axiomatic set of vectors.
 for
  v1 = q(**v1**)•u(**v1**)
  v2 = q(**v2**)•u(**v2**)
  angle(1,2), the angle between **u(v1)** and **u(v2)**
  un(v1,v2), a direction orthogonal
      to **v1** and **v2**
 then
  v1 ∧ **v2** ≡ q(**v1**)∗q(**v2**)∗sin(angle(1,2))∗**un(v1,v2)**

                                    end of definition

Multiplication2 is called the **cross product**. To avoid confusion with the other multiplications in the vector set it is symbolized with '∧'. As you can see

$$\wedge: V \times V \rightarrow V.$$

Newton: "The cross product depends not only on the angle between the two vectors, but also on an orthogonal direction. We must have
    un • **v1** = 0
and
    un • **v2** = 0
Does such a vector exist?"

Breton: "We need only
    un • **u(v1)** = 0
and
    un • **u(v2)** = 0
so we need only consider the unit sphere."

Newton: "Since all directions are part of our vector set, we can find one which is orthogonal to **u(v1)**."

Breton: "The two vectors, **v1** and **v2**, can be used to define a plane. The existence of the plane implies a direction orthogonal to it which we call **un**. Any of the vectors of this plane are orthogonal to **un**, and so we call this plane **orthogonal** to **un**."

Einstein, objecting: "When you talk about planes you are implying Euclidean geometry."

Breton: "Not really. We are only dealing with the axiomatic plus operator of the vector set. By 'plane' in this context I mean only a set P = {q1***v1** + q2***v2**, q1 and q2 in Q}. The vector plane would have to be further specified to make it a Euclidean plane."

Einstein: "A diagram would help here."

So Breton quickly sketched the following diagram to illustrate the orthogonal planes.

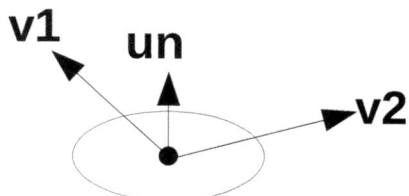

Breton: "Imagine the disk a unit circle viewed from the side. The circle is a great circle from a unit sphere which you have to imagine. The two vectors **v1** and **v2** lie in a plane which also contains the unit circle. The unit vector **un** is orthogonal to the unit circle and so orthogonal to both **v1** and **v2**."

Newton, conclusively: "So there *does* exist a direction which is orthogonal to both **v1** and **v2**."

Einstein, still objecting: "If **un** is orthogonal to **v1** then so also is –**un**. There are thus *two* orthogonal vectors, in opposite directions. So, Breton, your definition is flawed. You have narrowed the possibilities, but for an adequate definition you would have to narrow the possibilities to only one."

Breton: "True enough. Notice that the definition is only an *initial* definition. A final definition will be forthcoming."

Einstein: "Promises, promises, always promises."

Breton: "Which will be kept in due time. The initial definition of the cross product has narrowed the possibilities from an infinite number to just two. With a little patience we will finally narrow it to just one. But even with this initial definition we can come to some conclusions. What is **v1** ∧ **v1**?"

Newton: "Since sin(angle) = 0, **v1** ∧ **v1** = **0**."

Breton: "And how about **v1** ∧ **0**?"

Newton: "Since q(**0**) = 0, **v1** ∧ **0** = **0** also."

Breton: "Which also holds for **0** ∧ **v1**. And if **v1** • **v2** = 0?"

Newton: "Since sin(angle) = 1,
$$\mathbf{v1} \wedge \mathbf{v2} = q(\mathbf{v1})*q(\mathbf{v2})*\mathrm{un}(\mathbf{v1},\mathbf{v2}).\text{"}$$

Einstein: "Again an ambiguous result."

Breton, plodding forward patiently: "Which will be resolved anon. Notice when the value of the inner product is a minimum, the cross product has its maximum length. Conversely when $\mathbf{v1} \wedge \mathbf{v1} = 0$,
$\mathbf{v1} \bullet \mathbf{v1} = q(\mathbf{v1})*q(\mathbf{v1})$ attains its maximum value."

Einstein: "But when $\mathbf{v1} \wedge \mathbf{0} = \mathbf{0}$, then $\mathbf{v1} \bullet \mathbf{0} = 0$."

Newton, continuing: "We found the inner product interesting, but what possible interest can we expect from the cross product?"

Breton: "Look at the diagram again. If **v1** and **v2** are swirling, then an effect could be produced in the orthogonal direction. So to investigate the motion of propellers, we might find the cross product useful."

Einstein: "And for electricity as well."

Breton: "Interesting prospects, don't you think Newton? But let us focus again on our mathematical aim of defining an algebra for the axiomatic set of vectors. How about the third multiplication?"

Outer Product

Einstein: "Again we need a definition."

Breton: "Agreed. The way forward has become easier. Let me offer

> **Definition** (outer product)
> Given
> **v1, v2, v3** vectors in the axiomatic set of vectors.
>
> $\mathbf{v3} \bullet [\mathbf{v1} * \mathbf{v2}] \equiv (\mathbf{v3} \bullet \mathbf{v1}) * \mathbf{v2}$
>
> end of definition

As you can see the transformation [**v1** * **v2**] transforms the vector **v3** into a scaled vector parallel to **v2**. Now tell me: Is the transformation [**v1** * **v2**] identical with [**v2** * **v1**]?"

Einstein: "Of course not! The transformation [**v1** * **v2**] operates to produce a vector parallel to **v2** while [**v2** * **v1**] operates to produce a vector parallel to **v1**."

Newton: "Then although
$$\mathbf{v1} \cdot \mathbf{v2} = \mathbf{v2} \cdot \mathbf{v1}$$
still
$$[\mathbf{v1} * \mathbf{v2}] \neq [\mathbf{v2} * \mathbf{v1}].$$

Breton: "Which confirms the difference between these multiplications. In summary
$$\mathbf{v1} \cdot \mathbf{v2} = \mathbf{v2} \cdot \mathbf{v1}$$
$$\mathbf{v1} \wedge \mathbf{v2} = -\mathbf{v2} \wedge \mathbf{v1}$$
$$[\mathbf{v1} * \mathbf{v2}] \neq [\mathbf{v2} * \mathbf{v1}].$$
Suppose **v3** and **v1** are unit vectors. What would be the result?"

Newton: "Then
$$\mathbf{uv3} \cdot [\mathbf{uv1} * \mathbf{v2}] = \cos(\text{angle}(3,1)) * \mathbf{v2}$$
where cos(angle(3,1)) is the angle formed by **v3** and **v1**."

Breton: "What would be the result if the angle were 0?"

Newton: "Then cos(angle(3,1)) = 1, so
$$\mathbf{uv1} \cdot [\mathbf{uv1} * \mathbf{v2}] = \mathbf{v2}.$$

Breton: "Suppose **v3** and **v1** were orthogonal vectors."

Einstein: "Then cos(angle(3,1)) = 0, so
$$\mathbf{v3} \cdot [\mathbf{v1} * \mathbf{v2}] = \mathbf{0}.$$

Breton: "And if the angle equals 180 degrees?"

Einstein: "Then cos(angle) = –1, so
$$\mathbf{v3} \cdot [\mathbf{v1} * \mathbf{v2}] = -\mathbf{v2}.$$

Breton: "For given vectors, even though the multiplications are distinct, a certain symmetry appears in their ranges. Here let me illustrate by this table.

PRODUCT	RANGE
v1 · **v2**	q(**v1**)*q(**v2**) to – q(**v1**)*q(**v2**)
v1 ∧ **v2**	q(**v1**)*q(**v2**)*un to – q(**v1**)*q(**v2**)*un
v3 · [**v1** * **v2**]	q(**v1**)*q(**v3**)***v2** to – q(**v1**)*q(**v3**)***v2**

Einstein, reflecting: "Since the outer product transforms one vector into another, perhaps the cross product, which also yields a vector different from each multiplicand, can also be expressed as a transformation."

Newton, concerned about becoming defocused: "Before we wander off, let's stick to the trail of outer products."

Breton: " Can these multiplications be combined with addition? In Q, for instance, $2*(3+4) = (2*3) +(2*4)$. Is something like this valid for the set of vectors?"

Sums of inner products

Einstein, joining in enthusiastically to alleviate somewhat a suspicion that his contribution to the conversation was devolving into a pall of negativism: "Let me pose the question:
Does **v1** • **v2** + **v1** • **v3** equal **v1** • (**v2**+**v3**)?

The problem consists in the addition of two quotient numbers. Let angle(1,2) be the angle between **v1** and **v2**; let angle(1,3) be the angle between **v1** and **v3**. Then
v1 • **v2** = q(**v1**)*q(**v2**)*(**uv1**•**uv2**)
   = q(**v1**)*q(**v2**)*cos(angle(1,2))
v1 • **v3** = q(**v1**)*q(**v3**)*cos(angle(1,3))
v1 • **v2** + **v1** • **v3**
   = q(**v1**)*q(**v2**)*(**uv1**•**uv2**)
      + q(**v1**)*q(**v3**)*(**uv1**•**uv3**)
   = q(**v1**)*(q(**v2**)*(**uv1**•**uv2**) + q(**v3**)*(**uv1**•**uv3**)).

Breton appreciatively: "What can we say about
v1 • (**v2**+**v3**)?

Einstein: "That's not hard. As before, let
v2 = q(**v2**)•u(**v2**)
v3 = q(**v3**)•u(**v3**)
v2+**v3** = q(**v2**)•u(**v2**) + q(**v3**)•u(**v3**)
Then
v1 • (**v2**+**v3**)
   = **v1**•(q(**v2**+**v3**)•u(**v2**+**v3**).
   = q(**v1**)***uv1**•(q(**v2**)***uv2**) + q(**v3**)***uv3**)).

Breton: "So does **v1** • (**v2**+**v3**) = **v1**•**v2** + **v1**•**v3**?"

Einstein: "The formulas are almost the same."

Breton: "But not exactly. The would be equal if
(q(**v2**)*(**uv1**•**uv2**) + q(**v3**)*(**uv1**•**uv3**)
   = **u**(**v1**) • (q(**v2**)•u(**v2**)+q(**v3**)•u(**v3**)).

Einstein, recalling the earlier discussion on inner products: "Could they possible be equal if not identical?"

Breton: "An intriguing possibility. Let us take some examples.
If **v2** or **v3** = 0, say **v2** = 0, then
   v1•(**v2**+**v3**) = **v1**•**v3**
and   (**v1**•**v2**) + (**v1**•**v3**) = **v1**•**v3**;
likewise for **v3** = **0**.
  If **v2** = **v3**
      v1•(**v2**+**v3**) = **v1**•(**v2**+**v2**) = **v1**•(2***v2**)
         = 2•(**v1**•**v2**)

and $(v1 \cdot v2) + (v1 \cdot v3) = (v1 \cdot v2) + (v1 \cdot v2)$
$= 2*(v1 \cdot v2)$

If $v2 = -v3$
$$v1 \cdot (v2+v3) = v1 \cdot 0$$
$$= 0$$
and $(v1 \cdot v2) + (v1 \cdot v3) = (v1 \cdot v2) - (v1 \cdot v2)$
$= 0$

If $(v1 \cdot v2) = 0$ and $v1 \cdot v3 = 0$
$$v1 \cdot (v2+v3) = v1 \cdot q(v2+v3)*(a*u(v2) + b*u(v3))$$
$$= a*q(v2+v3)*v1 \cdot u(v2)$$
$$+ b*q(v2+v3)*v1 \cdot u(v3)$$
$$= 0 + 0$$
and $(v1 \cdot v2) + (v1 \cdot v3) = 0 + 0.$"

Einstein: "Encouraging."

Breton: "But not a proof that in every instance that
$$v1 \cdot (v2+v3) = v1 \cdot v2 + v1 \cdot v3.$$
Would you like to try proving the proposition generally?"

Einstein. amicably: "Let's try together. We already know they would be equal if
$(q(v2)*(uv1 \cdot uv2) + q(v3)*(uv1 \cdot uv3)$
$$= u(v1) \cdot (q(v2+v3)*u(v2+v3)).$$"

Breton: "We already know
$$q(v2)*(u(v1) \cdot u(v2)) = u(v1) \cdot (q(v2)*u(v2)$$
and similarly for $v3$. So we need only be concerned with the sum."

Einstein: "And we know that from our discussion of translated vectors
$$q(v2+v3)*u(v2+v3) = q(v2)*u(v2) + q(v3)*u(v3).$$"
So
$u(v1) \cdot (q(v2+v3)*u(v2+v3)$
$$= u(v1) \cdot (q(v2)*u(v2) + q(v3)*u(v3))$$
$$= u(v1) \cdot (q(v2)*u(v2) + u(v1) \cdot (q(v3)*u(v3))$$
$$= q(v2)*(uv1 \cdot uv2) + q(v3)*(uv1 \cdot uv3).$$"

Breton: "Not so fast. The two expressions
$$u(v1) \cdot (q(v2)*u(v2) + q(v3)*u(v3))$$
and
$$u(v1) \cdot (q(v2)*u(v2) + u(v1) \cdot (q(v3)*u(v3))$$
are different. When you equate them you are already assuming true what you want to prove."

Einstein: "This is hard to visualize."

Breton: "Let's look at the geometric rendition. Some drawings may be helpful."

In a few minutes Breton handed his friends three sketches.

Breton: "This first sketch shows the two vectors, **v2** and **v3** and their sum lying in the same plane. The vector **v1** sticks up from the plane. The dotted lines show the orthogonals from from **v1** to **v2**, **v3**, and **v2+v3**. The orthogonals are related to inner products.

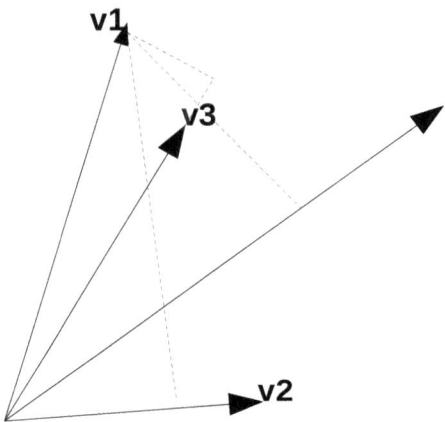

This next sketch show the area designated by **v1** • (**v2+v3**)

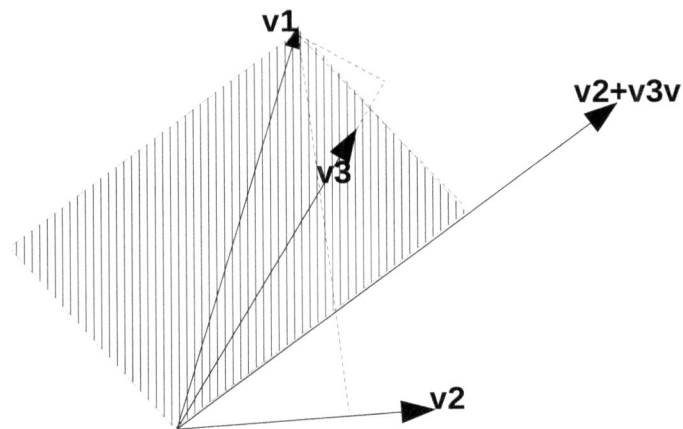

Geometrically, this area lies in the plane defined by **v1** and (**v2+v3**).

The next sketch shows the two areas **v1•v2** and **v1•v3**.

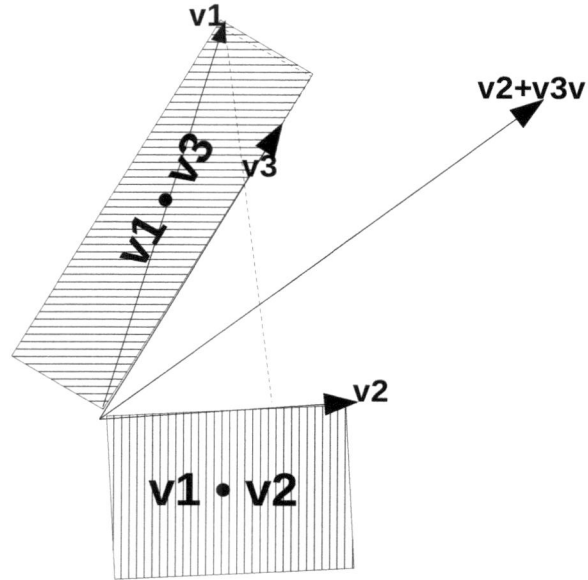

Breton: "The proposition would be proven if the first area equaled the sum of the latter two areas. It appears that a geometric proof would be even more difficult than an algebraic proof."

Einstein grudgingly: "So it does not follow that
v1•(v2+v3) equals **v1•v2 + v1•v3**."

Breton: "Patience. It only shows that we have not yet been able to prove their equality. When we have climbed somewhat higher I promise you a proof of their equality. For now, just accept it on my word."

Einstein still more grudgingly: "Promises, promises, still more promises."

Breton: "All of which will be kept anon. And I will make the same promise for an expression like **v1 ∧ (v2+v3)**. There are still other combinations of the multiplications to consider."

Combinations of Multiplications

Newton: "What? These multiplications can be combined?"

The scalar triple product

Breton: "Why not? The cross product of two vectors yields a vector which can then be a multiplicand of the inner product with a third vector to produce a quotient number. So isn't a combination like
$$\mathbf{v1} \cdot (\mathbf{v2} \wedge \mathbf{v3}) = q1$$
legitimate?"

Einstein: "Of course. Such combinations open interesting possibilities."

Breton: "So we can also consider combinations like
$$(\mathbf{v1} \wedge \mathbf{v2}) \cdot \mathbf{v3} = q2$$
and
$$\mathbf{v2} \cdot (\mathbf{v3} \wedge \mathbf{v1}) = q3.$$
Now I assert
$$q1 = q2 = q3.$$

Einstein: "Assert all you will, Breton. You will need to prove it before I accept it."

Breton: "It does seem astounding; you are right to question. If what I assert is true, we will have mounted a little higher up the mountain of our adventure from which we might expect to open up to a large panoramic vista.
  Let me start by making the proof a little easier. Defining
$$\mathbf{v1} \equiv qv1 \bullet \mathbf{uv1}$$
$$\mathbf{v2} \equiv qv2 \bullet \mathbf{uv2}$$
$$\mathbf{v3} \equiv qv3 \bullet \mathbf{uv3}$$
then
$\mathbf{v1} \cdot (\mathbf{v2} \wedge \mathbf{v3})$
    $= \mathbf{v1} \cdot (qv2 \bullet qv3 \bullet \sin(angle2,3) \bullet \mathbf{un23})$
    $= qv1 \bullet \mathbf{uv1} \cdot (qv2 \bullet qv3 \bullet \sin(angle2,3) \bullet \mathbf{un23})$
    $= qv1 \bullet qv2 \bullet qv3 \bullet \sin(angle2,3) \bullet \mathbf{uv1} \cdot \mathbf{un23}$
    $= qv1 \bullet qv2 \bullet qv3 \bullet \sin(angle2,3) \ast \cos(angle(\mathbf{v1},\mathbf{un23})).$
Likewise,
$\mathbf{v2} \cdot (\mathbf{v3} \wedge \mathbf{v1})$
    $= qv1 \bullet qv2 \bullet qv3 \bullet \sin(angle3,1) \ast \cos(angle(\mathbf{v2},\mathbf{un31}))$
$\mathbf{v3} \cdot (\mathbf{v1} \wedge \mathbf{v2})$
    $= qv1 \bullet qv2 \bullet qv3 \bullet \sin(angle1,2) \ast \cos(angle(\mathbf{v3},\mathbf{un12}))$
  The factor $qv1 \bullet qv2 \bullet qv3$ appears in all three equations where, as equal products, they form equal factors.
  So we need only consider whether
$$\sin(angle2,3) \ast \cos(angle(\mathbf{v1},\mathbf{un23}))$$
$$= \sin(angle3,1) \ast \cos(angle(\mathbf{v2},\mathbf{un31}))$$
$$= \sin(angle1,2) \ast \cos(angle(\mathbf{v3},\mathbf{un12})).$$

Einstein: "All these new definitions can be confusing."

Breton: "We are dealing with six different angles, so we need six different symbols. The complexity should be viewed as clarifying rather than confusing. If we don't create the symbols, our thinking would be very much impeded."

Newton, with a touch of impatience: "Breton, continue with your proof."

Breton: "Recall the rhombus which we used to defined the addition of vectors. What is its area?"

Einstein, impatiently: "What has this to do with the proof?"

Breton: "Patience, my dear Einstein."

Newton: "Everyone knows the area of a rhombus is the product of its base with its height."

Breton: "Not necessarily then, the product of its base with its *side*. Let me illustrate." With that Breton drew the following illustration.

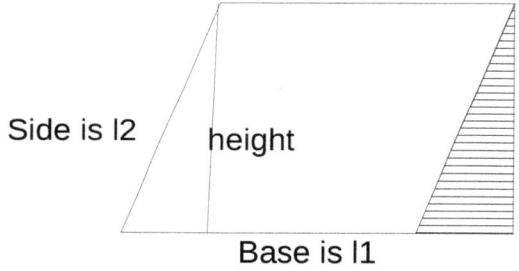

Base is l1

Area of a rhombus

Breton: "The area of the rhombus equals the product of its base and height because one can translate the triangle with the slanted side to the other side of the rhombus (the area indicated by the hatched triangle) to produce a square. Subtracting the area of one triangle and replacing with the area of an equal triangle does not change the value of the rhombus' area. The reconstructed area is then a square whose area is clearly the product of the length of its base with the length of its height.

Now consider the angle between the base and the side. The length of the height is then the length of the side times the sine of that angle. Let us call the length of the base, l1, and the length of the side, l2. The area of the rhombus can then be calculated as
$$\text{area} = l1 * l2 * \sin(\text{angle}(l1, l2)).$$

Newton: "A similar conclusion could be reached for rhomboids also."

Breton: "Certainly."

Einstein, questioning: "You have assigned base and side arbitrarily. If you exchanged them, might you not come to a different result?"

Newton: "I'm beginning to see how this argument could lead to proving your contention, Breton. But show us how it makes no difference which side is considered the base."

Breton: "It's not clear from the illustration? All right, let's switch the sides. Here is a second illustration,

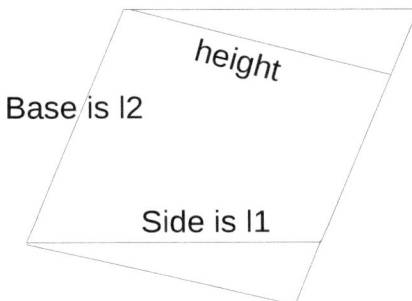

So you see the first side has become a new base with a different height, but the same area. So we can conclude
$$\text{area} = l1*l2*\sin(\text{angle}(l1,l2))$$
$$= l2*l1*\sin(\text{angle}(l2,l1)).$$
So we can further conclude
$$\sin(\text{angle}(l1,l2)) = \sin(\text{angle}(l2,l1)).$$

Newton: "What does that mean for vectors?"

Breton: "Let the sides and their directions be represented as vectors, **v1** and **v2**. Do you remember the definition of the cross product."

Newton: "Of course.
$$\mathbf{v1} \wedge \mathbf{v2} \equiv q(\mathbf{v1})*q(\mathbf{v2})*\sin(\text{angle}(1,2))*\mathbf{un}(\mathbf{v1},\mathbf{v2}).$$

Breton: "So look at the scalar part of the product. Isn't is just the area of a rhomboid?"

Newton: "Yes, I see that. Still, the cross product is a vector and not a scalar."

Breton: "Or we can consider the cross product a *vector* area, whose direction is orthogonal to the plane of its two vectors, and whose length equals the area of the rhomboid defined by the two vectors."

Einstein, mockingly: "So length equals area?"

Breton: "I stand corrected. I should have said 'whose *magnitude* equals the area of the rhomboid.'"

Einstein: "Much better. So just as the arithmetic product of two lengths is a scalar area, the cross product of two vectors is a vector area."

Newton: "Wonderful. The language of vectors subsumes ordinary arithmetic and even surpasses it."

Breton: "In this instance we now can consider areas as scalars or as vectors. Vectorial language is like singing a song rather than just reading the score."

Einstein: "Remarkable and surprising, but enough of metaphors! We know how to calculate the arithmetical area of a rhomboid and that it makes no difference which side is taken as base, but for vectorial areas, how do we know that either option has the same direction?"

Breton: "That's easy enough. Since the two vectors define a plane, any vector orthogonal to both vectors will be orthogonal to the plane, and thus be parallel vectors. The *order* of the vectors, however, becomes important, since a reversed order will produce a negatively parallel vector."

Einstein: "Breton, you still have not proven your assertion. How do vectorial areas help with your proof?"

Breton: "Consider now the volume of parallelepipeds."

Einstein: "Rhomboids, parallelepipeds! Just define a parallelepiped. Be like Newton who was good enough to define rhomboids."

Breton: "A parallelepiped is a six sided solid mathematical object, each side of which is a rhomboid which has a similar side opposite and parallel to it. Here I'll draw and illustration."

With that Breton drew the following which he presented to his

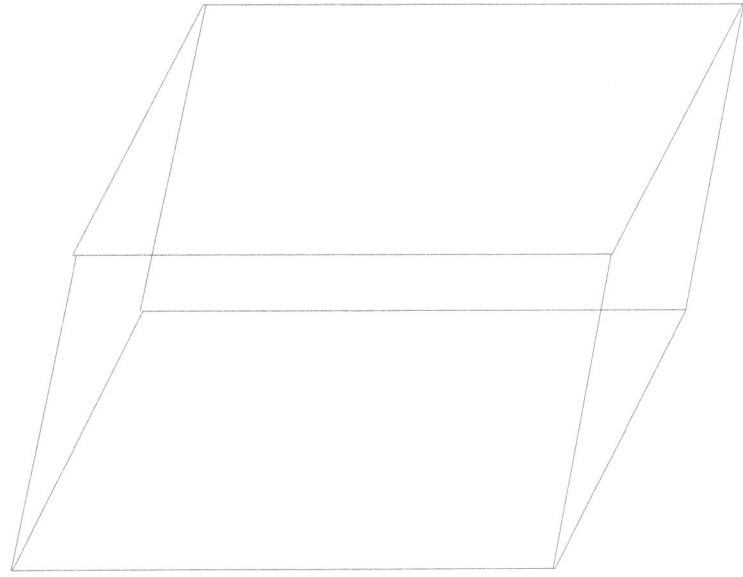

Parallelepiped

Newton: "So a parallelepiped is just an extension of rhomboids to three dimensions. Since each surface of the parallelepiped is a rhomboid we now know how to calculate its surface area."

Einstein: "So a parallelepiped is just a box."

Newton: "Which may be scrunched up a bit."

Breton: "We were able to calculate the area of a rhomboid from the knowledge of its sides. Because of parallelism we needed to know only two different sides.
 Now parallelpipeds have a volume. How can we calculate its volume?"

Newton: "For a rectangular parallelepiped, the answer is its base area times its height."

Breton: "Since the area of the base for the rectangular case is just the product of the length of its sides, the volume of the rectangular parallelepiped is just the product of the length of its three edges. More generally, if the base area is a rhomboid, while the remaining edge is perpendicular to the base, the volume of the parallelepiped would be the area of the base rhomboid times the length of the remaining side."

Einstein: "How about the general case where the remaining side is canted in an arbitrary direction with respect to the base."

Breton: "So let us go vectorial. Let each of the three non-parallel edges be designated **v1**, **v2**, and **v3**. Further let **v1** and **v2** be associated with the base area. Then the height of the parallelepiped is associated with **v3**. Let **un** be a direction orthogonal to the base vectorial area. The length of the height is thus
$$\text{length of height} = q(\mathbf{v3}) * \cos(\mathbf{uv3}, \mathbf{un}(\mathbf{v1},\mathbf{v2})).$$

Einstein: "Stay vectorial. Don't you see
$$q(\mathbf{v3}) * \cos(\mathbf{uv3}, \mathbf{un}) = \mathbf{v3} \cdot \mathbf{un}.$$

Breton: "Thank you, Einstein. Therefore,
volume = area of base*(length of height)
$$= q(\mathbf{v1})*q(\mathbf{v2})*\sin(\text{angle}(\mathbf{v1},\mathbf{v2}))$$
$$*q(\mathbf{v3})*(\mathbf{uv3} \cdot \mathbf{un}(\mathbf{v1},\mathbf{v2}))$$
Now consider
$$(\mathbf{v1} \wedge \mathbf{v2}) \cdot \mathbf{v3} = q(\mathbf{v1})*q(\mathbf{v2})*\sin(\text{angle}(\mathbf{v1},\mathbf{v2}))$$
$$*\mathbf{un}(\mathbf{v1},\mathbf{v2}) \cdot q(\mathbf{v1})*\mathbf{uv3}$$
$$= q(\mathbf{v1})*q(\mathbf{v2})*\sin(\text{angle}(\mathbf{v1},\mathbf{v2}))$$
$$*q(\mathbf{v3})*(\mathbf{un}(\mathbf{v1},\mathbf{v2}) \cdot \mathbf{uv3}).$$
What do you conclude?"

Newton: "Since **un**(**v1**,**v2**) and **un** are both directions orthogonal to the base, the equations are the same."

Breton: "Not the same, but equal. So $(\mathbf{v1} \wedge \mathbf{v2}) \cdot \mathbf{v3}$ equals the volume of a paralellepiped formed by the three vectors."

Einstein: "Nice, but this does not fully prove your assertion. You must further show that any combination of the vectors yields the same result."

Breton: "Fair enough. First do you agree
$$(\mathbf{v1} \wedge \mathbf{v2}) \cdot \mathbf{v3} = \mathbf{v3} \cdot (\mathbf{v1} \wedge \mathbf{v2})?$$

Einstein: "Certainly. As we saw earlier, inner products commute."

Breton: "So now I must show
$$(\mathbf{v1} \wedge \mathbf{v2}) \cdot \mathbf{v3} = (\mathbf{v2} \wedge \mathbf{v3}) \cdot \mathbf{v1}.$$

Both Newton and Einstein lean forward eagerly.

Breton: "Now $(\mathbf{v2} \wedge \mathbf{v3}) \cdot \mathbf{v1}$ corresponds to a different side of the parallelepiped with a different height. But since it is the same parallelepiped, tell me does it have the same volume?"

Einstein: "Yes."

Breton: "So. even though the multiplicative factors are different,
$$(\mathbf{v1} \wedge \mathbf{v2}) \cdot \mathbf{v3} = (\mathbf{v2} \wedge \mathbf{v3}) \cdot \mathbf{v1}$$
as volumes."

Einstein, even more reluctantly: "Yes."

Breton, pressing his argument home: "And
$$(\mathbf{v1} \wedge \mathbf{v2}) \cdot \mathbf{v3} = (\mathbf{v2} \wedge \mathbf{v3}) \cdot \mathbf{v1}$$
$$(\mathbf{v1} \wedge \mathbf{v2}) \cdot \mathbf{v3} = (\mathbf{v3} \wedge \mathbf{v1}) \cdot \mathbf{v2}$$
$$(\mathbf{v1} \wedge \mathbf{v2}) \cdot \mathbf{v3} = \mathbf{v1} \cdot (\mathbf{v2} \wedge \mathbf{v3})$$
$$(\mathbf{v1} \wedge \mathbf{v2}) \cdot \mathbf{v3} = \mathbf{v2} \cdot (\mathbf{v3} \wedge \mathbf{v1})$$
Does all this finally prove my assertion?"

Newton: "Yes indeed! What you have shown is marvelous indeed. Vectorial multiplication marvelously comprehends arithmetical multiplication and greatly extends it. I begin to see how our intellectual vistas are being enlarged."

Einstein: "Not so fast. Since,
$$\mathbf{v} \cdot (-\mathbf{v3}) = - (\mathbf{v} \cdot \mathbf{v3})$$
$$\neq \mathbf{v} \cdot \mathbf{v3}$$
$$(\mathbf{v1} \wedge \mathbf{v2}) \cdot (-\mathbf{v3}) \neq (\mathbf{v2} \wedge \mathbf{v3}) \cdot \mathbf{v1}$$
What's going on here?"

Breton: "Very little gets by you Einstein. What is
$$(\mathbf{v1} \wedge \mathbf{v2}) \cdot (-\mathbf{v3})?$$

Einstein: "You tell me."

Breton: "If $(\mathbf{v1} \wedge \mathbf{v2}) \cdot (\mathbf{v3})$ is the volume of a parallelepiped, then $(\mathbf{v1} \wedge \mathbf{v2}) \cdot (-\mathbf{v3})$ is the volume of a *different* parallelepiped."

Einstein: "With a negative volume!"

Breton: "Just so. You bring up an important subject which we touched on yesterday. Consider arithmetic multiplication. If the area of a rectangle
$$\text{area} = 2*3 = 6$$
what is the area of a rectangle $2*(-3)$?"

Einstein: "How can it be -6?"

Breton: "Recall how yesterday we distinguished two conventions: positive definite and basic. If we insist that all areas are positive then we are insisting on the positive definite convention. If not, then we should use the basic convention which allows an area to be negative. The basic convention insists that the positive area differs from the negative area. We risk confusion when we mix the conventions."

Einstein: "So we are using the basic convention with these vectorial multiplications."

Breton: "Correct. The basic convention was assumed when we agreed that
$$(v1 \cdot (-v3)) = - v1 \cdot v3.$$

Einstein: "Why not use the positive definite convention?"

Newton, interjecting: "Then we would not be able to use negative numbers."

Breton: "No small restriction. In any case, we have assumed the basic convention."

Einstein: "Then the word 'volume' is misleading."

Breton: "Only from the aspect of the positive definite convention. Our expanded (basic) view allows negative areas as well as negative volumes. This comports well with the possibility that $(v1 \wedge v2) \cdot (v3)$ may be positive or negative.
 If you insist on the positive definite convention, then we shall have to consider only
$$\text{abs}((v1 \wedge v2) \cdot (v3)).$$

Newton: "The basic convention will do for me."

Einstein: "Then we must see
$$(v1 \wedge v2) \cdot (v3)$$
as different from
$$(v2 \wedge v1) \cdot (v3)$$
since the second is the negative of the first."

Breton: "Correct. Here's a little mnemonic to help associating which volumes are equal. All the equal volumes keep a cyclic order of the vectors. For instance, $(v1 \wedge v2) \cdot v3$ orders the vectors as 1,2,3. It has the same volume as $(v2 \wedge v3) \cdot v1$ which orders the vectors 2,3,1.
 The cyclic order has
$$1,2,3 \quad \rightarrow \quad 2,3,1 \quad \rightarrow \quad 3,1,2$$
Vectors so ordered have equal values."

Einstein: "And the opposite direction
$$3,2,1 \quad \rightarrow \quad 1,3,2 \quad \rightarrow \quad 2,1,3$$
would also tag equal values?"

Breton: "Check it out."

Newton: "I will. After a short pause. "Look, the rule works flawlessly."

Breton: "So long as the cyclic order is preserved, any arrangement of the vectors yields the same value. For this reason, the value is often referred to as **the scalar triple product**."

Einstein: "Even though each of the products individually are different, the all have the same value. This is a result I find hard to stomach."

Breton: "We have here another instance of the important distinction between the meaning of the word 'is' and the meaning of the word 'equals'. It is just loose thinking to conflate the two. We could all agree that
$$7+3=4+6=5+5=10$$
even though {7+3} *is* not {4+6} which *is* not {5+5}."

Newton: "Yesterday, you insisted on the same distinction. I agreed then, but now realize better how the difference between 'is' and 'equals' is rooted deeply in our efforts to think correctly."

Breton: "You remind me now of another aspect of triple products which reflects a conclusion reached yesterday. If the vector **v** has units, say L, the inner, vector, and outer products we have defined should have units, L∗L, and the triple product units L∗L∗L. In our development taking L as length, then L∗L would denote an area and L∗L∗L a volume."

Newton: "Exactly as we determined."

Breton: "So vectorial algebra fits nicely, at least in this aspect, with our quest for Theoretical Physics."

Newton: "This is becoming intellectually satisfying."

Breton: "We have mounted a little further up our mountain. Let's turn to view the panorama. The vectors, **v1**, **v2**, and **v3** may have any physical units. For instance we might considered
$$(\mathbf{f1} \wedge \mathbf{f2}) \cdot (\mathbf{v3})$$
where **f1** and **f2** are forces and **v3** a velocity. Then we would know immediately
$$(\mathbf{f1} \wedge \mathbf{f2}) \cdot (\mathbf{v3}) = (\mathbf{v3} \wedge \mathbf{f1}) \cdot (\mathbf{f2})$$
$$= -(\mathbf{f1} \wedge \mathbf{v3}) \cdot (\mathbf{f2})$$
although measuring and calculating the involved variables might be difficult."

Newton, in amazement: "Instead of fashioning only one more intellectual idea for our explorations, we have now a whole warehouse of interesting intellectual tools."

Breton: "I find it pitiful that so much of modern physics is explained merely in terms of scalars. Vectorial explanations offer the prospect of so much more insight. For instance, the idea of area as a scalar has dominated our thinking, but the idea of area can be expanded as a vector, and perhaps even as a transformation. How might these expanded ideas of area enlighten our thinking?"

Newton: "Our metaphorical mountain offers more challenges than we first foresaw."

With that, Newton rose from his chair and took down the picture which he had framed during the previous day's conversation.

Breton: "We have extended ourselves. I think it time to regroup and consider our next steps.

Newton:"Time for lunch."

Einstein: "I agree. If we climb too fast we can slip and fall."

So the three friends rose from their Windsors, exited from their clubhouse to the same restaurant as yesterday where they relaxed over a savory meal.

Refreshed they returned to their Windsors with renewed vigor for the adventure.

Einstein began: "Newton, would you kindly summarize the morning's conversation for us."

Newton: "Let me start from the axioms themselves. We are given a field of scalar numbers (taken as Q, the quotient numbers), an axiomatic set of vectors, and two operators (addition and scalar multiplication) which act on any vector to produce another vector. Let me symbolize them as
$$+: V1 \times V2 \to V3$$
and
$$*: Q \times V1 \to V2.$$

Breton: "You show the sets involved nicely, but please show the symbolism for the action of the operators on individual elements of the vector set, **V**."

Newton: "Certainly
$$v1+v2 = v3$$
and
$$q*v1 = v2.$$
The axioms stipulate that these equations are always valid.

The vector set, axiomatically, does not define multiplication of vectors themselves. So of itself, it is not a field. Breton proposed expanding the vector set to include some other operators, like multiplication and so try to construct a field on the foundation of the axiomatic set of vectors. We defined three such possible multiplications: the inner, the cross, and the outer products."

Breton: "Would you construct a table showing what we have so far accomplished."

Newton: "Gladly."

With that Newton set to work and soon produced the following table which he passed to this two friends.

Axiomatic	Comments
$v1+v2 = v3$	closure
$q*v1 = v2$	Scalar multiply
$v1+(v2+v3) = (v1+v2)+v3$	association
Defined: two at a time	
$v1 \bullet v2 = v2 \bullet v1$	Inner product
$b*v1 \bullet c*v2 = b*c*(v1 \bullet v2)$	
$abs(v1 \bullet v2) \le abs(v1) * abs(v2)$	
$v1 \wedge v2 = -(v2 \wedge v1) = ((-v2) \wedge v1) = (v2 \wedge (-v1))$	cross product
$v1 \wedge v1 = 0$	
$(b*v1) \wedge (c*v2) = b*c*(v1 \wedge v2)$	
$abs(v1 \wedge v2) \le abs(v1)*abs(v2)$	
$v1 \bullet (v1 \wedge v2) = v2 \bullet (v1 \wedge v2) = 0$	
$(b*v1)*(c*v2) = b*c*(v1*v2)$	
$(abs(v1)*abs(v2))^2 = (abs(v1 \wedge v2))^2 + (abs(v1 \bullet v2))^2$	

Defined: three at at time	
v1•(**v2**∧**v3**) = **v2**•(**v3**∧**v1**) = **v3**•(**v1**∧**v2**) = (**v1**∧**v2**)•**v3** = (**v2**∧**v3**)•**v1** = (**v3**∧**v1**)•**v2**	Scalar triple product
v1•(**v2**∗**v3**) = (**v1**•**v2**)∗**v3**	transformation

Einstein: "Your list is impressive, but you have added some unproven equations."

Breton: "As usual Einstein, little escapes your notice. Newton, I see you have added only three such equations. Please tell us why and more importantly prove those assertions."

Newton: "All three equations refer to absolute values. Consider first
$$\text{abs}(\mathbf{v1}\cdot\mathbf{v2}) \le \text{abs}(\mathbf{v1}) * \text{abs}(\mathbf{v2}).$$
If we use basic convention
$$\text{abs}(\mathbf{v1}) = \text{abs}(qv1)$$
$$\text{abs}(\mathbf{v2}) = \text{abs}(qv2)$$
$$\text{abs}(\mathbf{v1}\cdot\mathbf{v2}) = \text{abs}(qv1*qv2*\cos(\text{angle}))$$
The result follows since $\text{abs}(\cos(\text{angle})) \le 1$.
  The second such equation
$$\text{abs}(\mathbf{v1}\wedge\mathbf{v2}) \le \text{abs}(\mathbf{v1}) * \text{abs}(\mathbf{v2})$$
follows almost immediately since
$$\text{abs}(\mathbf{v1}\wedge\mathbf{v2}) = \text{abs}(qv1*qv2*\sin(\text{angle}))$$
since again $\text{abs}(\sin(\text{angle})) \le 1$.
  The third such equation
$$(\text{abs}(\mathbf{v1})*\text{abs}(\mathbf{v2}))^2 = (\text{abs}(\mathbf{v1}\wedge\mathbf{v2}))^2 + (\text{abs}(\mathbf{v1}\cdot\mathbf{v2}))^2$$
follows closely.
Note
$$(\text{abs}(\mathbf{v1})*\text{abs}(\mathbf{v2}))^2 = (qv1*qv2)^2$$
$$\text{abs}(\mathbf{v1}\wedge\mathbf{v2}))^2 = (qv1*qv2*\sin(\text{angle}))^2$$
$$\text{abs}(\mathbf{v1}\cdot\mathbf{v2}))^2 = (qv1*qv2*\cos(\text{angle}))^2$$
so
$(\text{abs}(\mathbf{v1}\wedge\mathbf{v2}))^2 + (\text{abs}(\mathbf{v1}\cdot\mathbf{v2}))^2$
$$= (qv1*qv2)^2*(\sin^2(\text{angle}) + *\cos^2(\text{angle}))$$
$$= (qv1*qv2)^2.$$
$$= (\text{abs}(\mathbf{v1})*\text{abs}(\mathbf{v2}))^2.$$

Breton: "So this third equation simply rests on the identity
$$\sin^2(\text{angle}) + *\cos^2(\text{angle}) = 1.$$

Einstein: "Good, but it seems to me other combinations of three vectors are possible. For instance, Newton should add (**v1**•**v2**)∗**v3** to his table."

Newton: "Einstein, you're right. If **v1**•(**v2**∗**v3**) can make the list why not **v1**∧(**v2**∧**v3**)?"

Breton: "Agreed, but while $\mathbf{v1}\bullet(\mathbf{v2}*\mathbf{v3})$ is defined from the outer product what does $\mathbf{v1}\wedge(\mathbf{v2}\wedge\mathbf{v3})$ equal?"

Newton: "Let's not go too fast. I need equations to enter into my table."

Einstein: "From the definition of outer product we know
$$\mathbf{v1}\bullet(\mathbf{v2}*\mathbf{v3}) = (\mathbf{v1}\bullet\mathbf{v2})*\mathbf{v3}.$$

Breton: "We also know from in inner product that
$$\mathbf{v1}\bullet\mathbf{v2} = \mathbf{v2}\bullet\mathbf{v1}$$
so that
$$\mathbf{v1}\bullet(\mathbf{v2}*\mathbf{v3}) = (\mathbf{v2}\bullet\mathbf{v1})*\mathbf{v3}.$$

Einstein: "And again from the outer product
$$(\mathbf{v2}\bullet\mathbf{v1})*\mathbf{v3} = (\mathbf{v2}\bullet(\mathbf{v1}*\mathbf{v3}).$$

Newton: "Good. I'll add to my table
$$\begin{aligned}\mathbf{v1}\bullet(\mathbf{v2}*\mathbf{v3}) &= (\mathbf{v1}\bullet\mathbf{v2})*\mathbf{v3}\\ &= (\mathbf{v2}\bullet\mathbf{v1})*\mathbf{v3}\\ &= \mathbf{v2}\bullet(\mathbf{v1}*\mathbf{v3}).\end{aligned}$$

After a short interval in which Newton updated his table, he turned to his friends and asked: "What now?"

The Origin

Breton: "Let's consider the zero vector which is provided axiomatically in the axiomatic set of vectors. Since it is a vector, what is its magnitude?"

Einstein: "That's easy. The zero vector has zero length."

Breton: "By which you must mean,
$$\mathbf{0} = 0*u(\mathbf{0})$$
where zero in the axiomatic set of vectors is not the same as zero in the underlying field of quotient numbers, Q."

Einstein: "Thank you for your precision."

Breton: "Now prove your assertion."

Einstein: "Of course, it's true."

Breton: "Which is merely your assertion. Why can't the zero vector have some other magnitude?"

Newton:"Why not?"

Einstein: "All right, let me try to prove something I already know is true. Where do I start?"

Breton: "You might try the axioms."

Einstein: "Of course, the axioms are taken as true. Where shall we start?"

Breton: "For any vector
$$1*\mathbf{v} + 1*\mathbf{v} = (1+1)*\mathbf{v} = 2*\mathbf{v}$$
$$1*\mathbf{v} - 1*\mathbf{v} = (1-1)*\mathbf{v} = 0*\mathbf{v}$$
Now what can you say about abs(**0**)?"

Einstein: "Since abs($\mathbf{v}$)≥0 for any vector
$$\text{abs}(\mathbf{0}) \geq 0.$$

Breton: "And is abs($\mathbf{v} + \mathbf{v}$)≥abs($\mathbf{v} - \mathbf{v}$)?"

Einstein: "Yes indeed. So for $\mathbf{v} = \mathbf{0}$
$$0 \leq \text{abs}(\mathbf{0}) = \text{abs}(\mathbf{v} - \mathbf{v}) \leq \text{abs}(\mathbf{v} + \mathbf{v}) \leq 2*\text{abs}(\mathbf{0})$$
For any non-zero value for abs(**0**) the inequalities would not hold. Only abs(**0**) = 0 works."

Newton: "So you have proven abs(**0**) = 0!"

Breton: "Now let me ask you another question. Since every vector can be written
$$\mathbf{v} = \text{abs}(\mathbf{v})*\mathbf{u}(\mathbf{v})$$
that is, the scalar product of a magnitude and a direction, what is the direction of **0**?"

Einstein: "The zero vector like the zero in the numbers is very special."

Breton: "Is it? Notice
$$\mathbf{0} = \text{abs}(\mathbf{0}) * \mathbf{u}(\mathbf{0})$$
$$= 0 * \mathbf{u}(\mathbf{0})$$
so it appears any direction will do for $\mathbf{u}(\mathbf{0})$."

Einstein: "Breton, you never change your rascally ways."

Breton: "If $\mathbf{u}(\mathbf{0})$ is any one of many possible directions, then **0** can be any one of many possible vectors."

Newton: "But still it *is* only one vector."

Einstein: "What a morass you have led us into, Breton. It has to be one of many possible unspecified vectors."

Newton: "In fact any one of an infinite number of vectors, since directions correspond to all the points on the unit sphere."

Breton: "Remember the axiom concerning **0**
$$\mathbf{v} = \mathbf{v} + \mathbf{0}?$$
So **0** acts as a reference for all vectors. This is true for Q as a vector set where the quotient number 0 acts as a reference; it is also true for *magnitudes* in **V**, as Einstein has just shown,
$$\text{abs}(\mathbf{v}) = \text{abs}(\mathbf{v}) + \text{abs}(\mathbf{0}).$$
So how can **0** serve as a reference for the direction of any vector?"

Newton: "Take any direction **u1**. The the subset $\{\mathbf{v}|\mathbf{v} = q*\mathbf{u1}$ for all $q\}$ is a line of vectors whose directions can all be referenced to **u1**."

Breton: "Splendid. So **u1** can be part of our answer."

Newton: "Next take **u2** any direction orthogonal to **u1**. The subset $\{\mathbf{v}|\mathbf{v}= q1*\mathbf{u1} + q2*\mathbf{u2}$ for all $q1$ and $q2\}$ is a plane of vectors whose directions can all be referenced to **u1** and **u2**."

Breton: "Splendid. So **u1** and **u2** can be part of our answer."

Newton: "Next take **u3** any direction orthogonal to **u1** and **u2**. The the subset $\{\mathbf{v}|\mathbf{v}= q1*\mathbf{u1} + q2*\mathbf{u2} + q3*\mathbf{u3}$ for all $q1$ and $q2$ and $q3\}$ is the set of all vectors each of which can be referenced to **u1** and **u2** and **u3**."

Einstein: "So in Newton's scheme we first choose arbitrarily one direction **u1** of all the directions of the sphere; next we choose arbitrarily a second direction **u2** from a great circle of the sphere orthogonal to **u1**; finally we no longer choose but accept the single direction **u3** which is orthogonal to both **u1** and **u2**.

 Any one of the three directions taken singly serves as a reference to an axiomatic subset of vectors in a line. Any two together serve as a reference to a subaxiomatic set of vectors in a plane. All three together serve as a reference to any vector."

Breton: "Including **0**?"

Einstein: "Yes. I like the balance between the choosing and the application."

Breton: "So now we have a reference for directions which though not derived from **0** will comport well with calling **0** the reference for all vectors, for both their magnitudes and directions. That is,
$$\mathbf{v} = q1*\mathbf{u1} + q2*\mathbf{u2} + q3*\mathbf{u3} + 0*\mathbf{u1} + 0*\mathbf{u2} + 0*\mathbf{u3}$$
specifying
$$\mathbf{v} = \mathbf{v} + \mathbf{0}.$$

Einstein: "How does this match
$$\mathbf{v} = q*\mathbf{u}(\mathbf{v})?$$

Breton: "The answer can be found using some of our previous results. The inner product

$$\mathbf{v} \bullet \mathbf{v} = q^2$$

for $\mathbf{v} = q*\mathbf{u}(\mathbf{v})$

and
$$\begin{aligned}
\mathbf{v} \bullet \mathbf{v} &= (q1*\mathbf{u1} + q2*\mathbf{u2} + q3*\mathbf{u3}) \\
&\quad \bullet (q1*\mathbf{u1} + q2*\mathbf{u2} + q3*\mathbf{u3}) \\
&= q1*\mathbf{u1} \bullet (q1*\mathbf{u1}) \\
&\quad + q1*\mathbf{u1} \bullet (q2*\mathbf{u2}) \\
&\quad + q1*\mathbf{u1} \bullet (q3*\mathbf{u3}) \\
&\quad + q2*\mathbf{u2} \bullet (q1*\mathbf{u1}) \\
&\quad + q2*\mathbf{u2} \bullet (q2*\mathbf{u2}) \\
&\quad + q2*\mathbf{u2} \bullet (q3*\mathbf{u3}) \\
&\quad + q31*\mathbf{u3} \bullet (q1*\mathbf{u1}) \\
&\quad + q3*\mathbf{u3} \bullet (q2*\mathbf{u2}) \\
&\quad + q3*\mathbf{u3} \bullet (q3*\mathbf{u3}) \\
&= q1*q1 + q2*q2 + q3*q3
\end{aligned}$$
since $\mathbf{ui} \bullet \mathbf{uj} = 0$ if i≠j and $\mathbf{ui} \bullet \mathbf{uj} = 1$ if i=j.

Newton: "So

$$q^2 = q1^2 + q2^2 + q3^2$$

Breton: "By specifying **0** this way, we have made it a reference for all vectors, both their magnitudes and directions."

Einstein: "So this new **0** is different from the axiomatic **0**!"

Breton: "To avoid ambiguity, let us call the new **0** the **origin** of our axiomatic set of vectors."

Newton: "The origin fits in with our previous discussion of direction and angles. One direction can be specified by three angles."

Breton: "There is something profound here, more profound than Mathematics or Theoretical Physics. It has been revealed that God (whom we know exists) is a Trinity, one God in three distinct Persons. It is not strange that his creation should show traces of its origin. We have such a trace here: one direction expressed as three distinct directions."

Einstein: "Are you saying you can prove God is a Trinity?"

Breton: "Of course not. While we have proved God's existence, what God *is* appears beyond our power of comprehension. But if God reveals himself as Trinity, then the world becomes more understandable."

Einstein: "We defined Physics as a science which deals with the world ais extended, moving, and forcing. The Trinity is none of these things. So God, the Trinity, is not physical, and so not the object of scrutiny by the science of Physics."

Breton: "I agree. God is like a frame around a picture. God gives meaning to the science of Physics, but is not the direct object of its study. Like a frame around a picture."

Newton: "A most interesting subject I agree, but not on our path up the mountain. Is there more about the origin?"

Breton: "Let's examine how a direction is expressed in terms of the origin's reference."

Einstein: "That's easy enough. Given the origin as described above, any direction
$$\mathbf{u}(\mathbf{v}) = q_1*\mathbf{u1} + q_2*\mathbf{u2} + q_3*\mathbf{u3})$$
for some quotient numbers, q_1, q_2, and q_3 where
$$\operatorname{sqrt}(q_1^2 + q_2^2 + q_3^2) = 1$$
from what we have just proven."

Breton: "For directions may I suggest replacing the symbol *q* with the symbol *c*."

Newton: "Why?"

Breton: "Because from a geometrical perspective, the c's are just directional cosines. It makes the relationship between direction and angles clear and exact."

Newton: "So vectors referred to an origin for **0** have two representations, one in terms of magnitude and direction, and a second in terms of the arbitrary coordinate system. There must exist relationships between the two representations."

Breton: "Let's examine them. Any vector
$$\mathbf{v} = \operatorname{abs}(\mathbf{v})*\mathbf{u}(\mathbf{v})$$
$$\mathbf{v} \equiv q*(c_1*\mathbf{u1} + c_2*\mathbf{u2} + c_3*\mathbf{u3})$$
$$\mathbf{v} \equiv q_1*\mathbf{u1} + q_2*\mathbf{u2} + q_3*\mathbf{u3}$$
$$\mathbf{v} = \mathbf{v} \bullet \mathbf{u1}*\mathbf{u1} + \mathbf{v} \bullet \mathbf{u2}*\mathbf{u2} + \mathbf{v} \bullet \mathbf{u3}*\mathbf{u3}$$
$$\mathbf{v} = \mathbf{v} \bullet (\mathbf{u1}*\mathbf{u1} + \mathbf{u2}*\mathbf{u2} + \mathbf{u3}*\mathbf{u3})$$
The first of these relationships we have from the axioms; the second is a definition of q as $\operatorname{abs}(\mathbf{v})$ for the magnitude of **v** with reference to the origin; the third defines three magnitudes in the origin's directions; the fourth equation relates these three magnitudes to the inner products with the given vector; the fifth equation factors the fourth equation and establishes
$$\mathbf{I} \equiv \mathbf{u1}*\mathbf{u1} + \mathbf{u2}*\mathbf{u2} + \mathbf{u3}*\mathbf{u3}$$
as the identity transformation.
It follows that
$$q_i = \mathbf{v} \bullet \mathbf{ui}$$
$$q = \operatorname{sqrt}(q_1^2 + q_2^2 + q_3^2)$$
$$c_i = q_i/q$$
The three c_i's are called directional cosines."

Einstein: "For a direction then
$$\text{abs}(\mathbf{u}(\mathbf{v})) = \sqrt{q_1^2 + q_2^2 + q_3^2} = 1$$

Newton: "Just as we asserted earlier. For any vector
$$\mathbf{v} = \text{abs}(\mathbf{v}) * \mathbf{u}(\mathbf{v})$$
$$= q_1 * \mathbf{u1} + q_2 * \mathbf{u2} + q_3 * \mathbf{u3}$$
$$= q * (c_1 * \mathbf{u1} + c_2 * \mathbf{u2} + c_3 * \mathbf{u3})$$
so
$$q_1 = q * c_1$$
$$q_2 = q * c_2$$
$$q_3 = q * c_3$$
So tell us why the c_i are called direction cosines."

Breton: "The equations give the reason
$$q_i = \mathbf{v} \bullet \mathbf{ui}$$
$$= q * \mathbf{u}(\mathbf{v}) \bullet \mathbf{ui}$$
so
$$q_i/q = \mathbf{u}(\mathbf{v}) \bullet \mathbf{ui}$$
$$= c_i$$
$$= \cos(\mathbf{u}(\mathbf{v}), \mathbf{ui})$$

Einstein: "That explains why the c_i are cosines, but why are they called *directional* cosines?

Breton: "The angles of the three cosines define the vector's direction. This diagram may help visualize what I am saying.

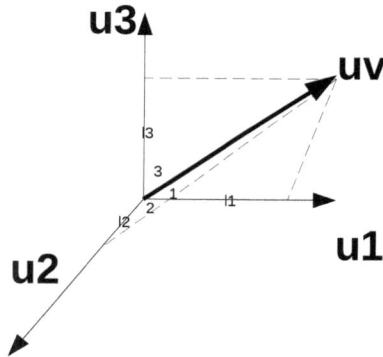

Breton: "Notice the origin's orientation, **u1**, **u2**, and **u3**, are perpendicular to each other. The direction of the vector is indicated by **uv**. Since **uv** is a direction, its length is one. The dashed lines are extend from the tip of **uv** to a point perpendicular to each of the origin's directions. You can see three angles marked 1, 2, and 3 between **uv** and **u1**, **u2**, and **u3**. While not well illustrated, the dashed lines form three right triangles each of which has q(**uv**) = 1 for its hypotenuse. The length of the adjacent side of each triangle is

labeled l1, l2, and l3. Now the cosine of each angle equals $li/q(\mathbf{uv})$ = li, i=1,2,3.
  Moreover
$$\mathbf{uv} = l1*\mathbf{u1} + l2*\mathbf{u2} + l3*\mathbf{u3}$$
$$= \cos(\text{angle}(1))*\mathbf{u1}$$
$$+\cos(\text{angle}(2))*\mathbf{u2}$$
$$+\cos(\text{angle}(3))*\mathbf{u3}$$
$$= \mathbf{uv}\bullet\mathbf{u1}*\mathbf{u1} + \mathbf{uv}\bullet\mathbf{u2}*\mathbf{u2} + \mathbf{uv}\bullet\mathbf{u3}*\mathbf{u3}$$

So this is why, my dear Einstein, the cosines are called *directional* cosines."

Newton, looking to reestablish the focus of the conversation: "How does representation in terms of the origin match with the vectorial operations."

Breton: "Easy enough. For
$$\mathbf{v1} = q1*\mathbf{uv1}$$
$$= q1*(c11*\mathbf{u1} + c12*\mathbf{u2} + c13*\mathbf{u3})$$
$$\mathbf{v2} = q2*\mathbf{uv2}$$
$$= q2*(c21*\mathbf{u1} + c22*\mathbf{u2} + c23*\mathbf{u3})$$
$$\mathbf{v1}+\mathbf{v2} = (q1*c11 + q2*c21)*\mathbf{u1}$$
$$+ (q1*c12 + q2*c22)*\mathbf{u2}$$
$$+ (q1*c13 + q2*c23)*\mathbf{u3}$$
$$\mathbf{v1}\bullet\mathbf{v2} = q1*q2*(c11*c21 + c12*c22 + c13*c23)$$
$$\mathbf{v1}\wedge\mathbf{v2} = q1*q2*((c12*c23 - c13*c22)*\mathbf{u1}$$
$$+ (c13*c21 - c11*c23)*\mathbf{u2}$$
$$+ (c11*c22 - c12*c21)*\mathbf{u3})$$
$$\mathbf{v1}*\mathbf{v2} = q1*q2*(c11*\mathbf{u1} + c12*\mathbf{u2} + c13*\mathbf{u3})$$
$$*(c21*\mathbf{u1} + c22*\mathbf{u2} + c23*\mathbf{u3}).$$
$$= q1*q2*(c11*\mathbf{u1}*(c21*\mathbf{u1} + c22*\mathbf{u2} + c23*\mathbf{u3})$$
$$+ c12*\mathbf{u2}*(c21*\mathbf{u1} + c22*\mathbf{u2} + c23*\mathbf{u3})$$
$$+ c13*\mathbf{u3})*(c21*\mathbf{u1} + c22*\mathbf{u2} + c23*\mathbf{u3})).$$
$$= q1*q2*(c11*c21*\mathbf{u1}*\mathbf{u1}$$
$$+ c11*c22*\mathbf{u1}*\mathbf{u2}$$
$$+ c11*c23*\mathbf{u1}*\mathbf{u3}$$
$$+ c12*c21*\mathbf{u2}*\mathbf{u1}$$
$$+ c12*c22*\mathbf{u2}*\mathbf{u2}$$
$$+ c12*c23*\mathbf{u2}*\mathbf{u3}$$
$$+ c13*c21*\mathbf{u3}*\mathbf{u1}$$
$$+ c13*c22*\mathbf{u3}*\mathbf{u2}$$
$$+ c13*c23*\mathbf{u3}*\mathbf{u3})$$

Einstein: "Wherever did you get $\mathbf{v1}\wedge\mathbf{v2}$?"

Breton: "I will now answer your question about the ambiguity in the cross product. Please follow these straight-forward substitutions and operations.

$\mathbf{v1} \wedge \mathbf{v2}$ = q1*q2*$\mathbf{uv1} \wedge \mathbf{uv2}$
     = q1*q2
        (c11$\mathbf{u1}$ $\wedge$(c21*$\mathbf{u1}$ + c22*$\mathbf{u2}$ + c23*$\mathbf{u3}$)
        + c12*$\mathbf{u2}$ $\wedge$(c21*$\mathbf{u1}$ + c22*$\mathbf{u2}$ + c23*$\mathbf{u3}$)
        + c13*$\mathbf{u3}$)$\wedge$(c21*$\mathbf{u1}$ + c22*$\mathbf{u2}$ + c23*$\mathbf{u3}$)
     = q1*q2*(c11*c21*$\mathbf{u1} \wedge \mathbf{u1}$
            + c11*c22*$\mathbf{u1} \wedge \mathbf{u2}$
            + c11*c23*$\mathbf{u1} \wedge \mathbf{u3}$)
            + c12*c21*$\mathbf{u2} \wedge \mathbf{u1}$
            + c12*c22*$\mathbf{u2} \wedge \mathbf{u2}$
            + c12*c23*$\mathbf{u2} \wedge \mathbf{u3}$)
            + c13*c21*$\mathbf{u3} \wedge \mathbf{u1}$
            + c13*c22*$\mathbf{u3} \wedge \mathbf{u2}$
            + c13*c23*$\mathbf{u3} \wedge \mathbf{u3}$)

Since the directions of the origin make up an orthogonal set, the following definitions resolve the ambiguity in the cross product.

$$\mathbf{u1} \wedge \mathbf{u2} = \mathbf{u3}$$
$$\mathbf{u2} \wedge \mathbf{u3} = \mathbf{u1}$$
$$\mathbf{u3} \wedge \mathbf{u1} = \mathbf{u2}$$

You can see that the cyclic arrangement I alluded to before is incorporated in these definitions.

Remembering that $\mathbf{ui} \wedge \mathbf{ui}$ = 0 and that $\mathbf{ui} \wedge \mathbf{uj}$ = $-\mathbf{uj} \wedge \mathbf{ui}$
$\mathbf{v1} \wedge \mathbf{v2}$ = q1*q2*(c11*c22*$\mathbf{u3}$
                − c11*c23*$\mathbf{u2}$
                − c12*c21*$\mathbf{u3}$
                + c12*c23*$\mathbf{u1}$
                + c13*c21*$\mathbf{u2}$
                − c13*c22*$\mathbf{u1}$)
     = q1*q2*(c12*c23*$\mathbf{u1}$
                − c13*c22*$\mathbf{u1}$)
                + c13*c21*$\mathbf{u2}$
                − c11*c23*$\mathbf{u2}$
                + c11*c22*$\mathbf{u3}$
                − c12*c21*$\mathbf{u3}$
     = q1*q2*((c12*c23 − c13*c22)*$\mathbf{u1}$)
                + (c13*c21 − c11*c23)*$\mathbf{u2}$
                + (c11*c22 − c12*c21)*$\mathbf{u3}$).

Newton, agreeably: "Just as you stated."

Einstein, stubbornly: "So the direction of the cross product depends on the choice of origin.

Breton: "Yes. This should not be too surprising. If you turn yourself upside down, what first faces up afterwards faces down."

Newton: "So the direction of the cross product which might have been defined completely arbitrarily at first, is finally defined in terms of an arbitrary origin. This must be why you refused to reduced the ambiguity earlier, despite Einstein's scoffing."

Einstein, defensively: "Breton, are you saying our algebra of vector sets then applies only to a particular choice of origin."

Breton: "That is a question which separates your illustrious ancestors. Isaac Newton held that one special location in the universe is an absolute location. His 'discoveries' depended on such an assumption. Albert Einstein disagreed. The origin of our algebra for the set of vectors may illumine the controversy. So let us put the question on our agenda, but first let us clear this path about vectorial operations referred to the origin a little more. Let me return to the sum of inner products. Now, Einstein, I will fulfill the promise I made earlier. I intend to prove that **(v1•v2) + (v1•v3)** equals **v1 • (v2+v3)**!"

Einstein: "And I will be scrutinizing every step of the proof."

Breton: "Good. Let me put the proof in form.

Proof: (distribution of an inner product)
Given

    u1,u2, and **u3** an orthogonal orientation of **V3**
    v1 = v11•**u1**+v12•**u2**+v13•**u3**
    v2 = v21•**u1**+v22•**u2**+v23•**u3**
    v3 = v31•**u1**+v32•**u2**+v33•**u3**

then

$$\mathbf{v1} \cdot (\mathbf{v2}+\mathbf{v3}) = (\mathbf{v1} \cdot \mathbf{v2}) + (\mathbf{v1} \cdot \mathbf{v3})$$

Proof:
v1•v2 = v11∗v21 + v12∗v22 + v13∗v23
v1•v3 = v11∗v31 + v12∗v32 + v13∗v33
v2+v3 = (v21+v31)•**u1**
          + (v22+v32)•**u2**
          + (v23+v33)•**u3**
v1 • (v2+v3) = v11∗(v21+v31)
              + v12∗(v22+v32)
              + v13∗(v23+v33)
Therefore,
v1 • (v2+v3) = v11∗v21+v11∗v31
                + v12∗v22+v12∗v32
                + v13∗v23+v13∗v33
           = v11∗v21 + v12∗v22 + v13∗v23
              + v11∗v31 + v12∗v32 + v13∗v33
           = **(v1•v2) + (v1•v3)**
                                      qed

Breton: "So what appeared as difficult from a geometric perspective has become much easier once the set of vectors has been given an orientation. In the geometric perspective we would have to prove two different areas equal a third. With a given orientation we need only do some arithmetic bookkeeping. Having now proved the proposition, we know the two areas do equal the third."

Einstein, objecting: "For one orientation. Would the same be true for a different set of coordinates or even for a different origin?"

Breton: "The proof does not rely on a specific orientation, but rather on any orientation. So in a different orientation, the same result would hold, though it might be expressed differently. For instance, in a different orientation the areas might be different, but still two of the areas would equal the third.
  Now let me settle a second promise, namely that
$$v1 \wedge (v2+v3) \text{ equals } v1 \wedge v2 + v1 \wedge v3.\text{"}$$

Sums of cross products

Newton: "May I attempt the proof?"

To which both Breton and Einstein nodded agreement.

Newton: "Let me first define
$$v1 \wedge v2 = q(v1)*q(v2)*\sin(\text{angle}(1,2))*un(v1,v2)$$
$$v1 \wedge v3 = q(v1)*q(v3)*\sin(\text{angle}(1,3))*un(v1,v3)$$
so
$v1 \wedge v2 + v1 \wedge v3$
$$= q(v1)*q(v2)*\sin(\text{angle2})*un(v1,v2)$$
$$+ q(v1)*q(v3)*\sin(\text{angle3})*un(v1,v3))$$
while
$v1 \wedge (v2+v3) = q(v1)*u(v1) \wedge (q(v2)*u(v2) + q(v3)*u(v3))$
$$= u(v1) \wedge (q(v1)*q(v2)*u(v2)$$
$$+ q(v1)*q(v3)*u(v3))$$
$$= q(v1)*q(v2)*u(v1) \wedge u(v2)$$
$$+ q(v1)*q(v3)*u(v1) \wedge u(v3))$$
$$= q(v1)*q(v2)*\sin(\text{angle}(1,2))*un(v1,v2)$$
$$+ q(v1)*q(v3)*\sin(\text{angle}(1,3))*un(v1,v3))$$
So they *are* equal."

Einstein: "You've fallen into the same error as we did with inner products. Your proof presupposes what you want to prove. When you claim
$u(v1) \wedge (q(v1)*q(v2)*u(v2) + q(v1)*q(v3)*u(v3))$
$$= q(v1)*q(v2)*u(v1) \wedge u(v2)$$
$$+ q(v1)*q(v3)*u(v1) \wedge u(v3))$$
you have just assumed what you wanted to prove!"

Breton: "Why not try with an assumed orientation?"

Newton: "How?"

Breton: "Remember
$$\mathbf{v1} \wedge \mathbf{v2} = q1*q2*((c12*c23 - c13*c22)*\mathbf{u1}$$
$$+ (c13*c21 - c11*c23)*\mathbf{u2}$$
$$+ (c11*c22 - c12*c21)*\mathbf{u3})$$
So do the bookkeeping."

Newton: "Let me assume an orientation specifying $\mathbf{u1}$, $\mathbf{u2}$, and $\mathbf{u3}$. Express
$\quad\quad\mathbf{v1}$ as $q1*(c11*\mathbf{u1} + c12*\mathbf{u2} + c13*\mathbf{u3}$
$\quad\quad\mathbf{v2}$ as $q2*(c21*\mathbf{u1} + c22*\mathbf{u2} + c23*\mathbf{u3}$
$\quad\quad\mathbf{v3}$ as $q3*(c31*\mathbf{u1} + c32*\mathbf{u2} + c33*\mathbf{u3}$
Then
$\mathbf{v1} \wedge \mathbf{v2} = q1*q2*((c12*c23 - c13*c22)*\mathbf{u1}$
$\quad\quad\quad\quad + (c13*c21 - c11*c23)*\mathbf{u2}$
$\quad\quad\quad\quad + (c11*c22 - c12*c21)*\mathbf{u3})$
$\mathbf{v1} \wedge \mathbf{v3} = q1*q3*((c12*c33 - c13*c32)*\mathbf{u1}$
$\quad\quad\quad\quad + (c13*c31 - c11*c33)*\mathbf{u2}$
$\quad\quad\quad\quad + (c11*c32 - c12*c31)*\mathbf{u3})$
So
$\mathbf{v1} \wedge \mathbf{v2} + \mathbf{v1} \wedge \mathbf{v3}$
$\quad = q1*q2*((c12*c23 - c13*c22)*\mathbf{u1}$
$\quad\quad\quad\quad + (c13*c21 - c11*c23)*\mathbf{u2}$
$\quad\quad\quad\quad + (c11*c22 - c12*c21)*\mathbf{u3})$
$\quad\quad\quad + q1*q3*((c12*c33 - c13*c32)*\mathbf{u1}$
$\quad\quad\quad\quad + (c13*c31 - c11*c33)*\mathbf{u2}$
$\quad\quad\quad\quad + (c11*c32 - c12*c31)*\mathbf{u3})$
$\quad = q1*q2*(c12*c23 - c13*c22)*\mathbf{u1}$
$\quad\quad + q1*q3*((c12*c33 - c13*c32)*\mathbf{u1}$
$\quad\quad + q1*q2*(c13*c21 - c11*c23)*\mathbf{u2}$
$\quad\quad + q1*q3*(c13*c31 - c11*c33)*\mathbf{u2}$
$\quad\quad + q1*q2*(c11*c22 - c12*c21)*\mathbf{u3}$
$\quad\quad + q1*q3*(c11*c32 - c12*c31)*\mathbf{u3}$
$\quad = (q1*q2*(c12*c23 - c13*c22)$
$\quad\quad\quad + q1*q3*(c12*c33 - c13*c32))*\mathbf{u1}$
$\quad\quad +(q1*q2*(c13*c21 - c11*c23)$
$\quad\quad\quad + q1*q3*(c13*c31 - c11*c33))*\mathbf{u2}$
$\quad\quad +(q1*q22*(c11*c22 - c12*c21)$
$\quad\quad\quad + q1*q3*(c11*c32 - c12*c31))*\mathbf{u3}$
$\quad = (q1*q2*c12*c23 - q1*q2*c13*c22$
$\quad\quad + q1*q3*c12*c33 - q1*q3*c13*c32)$
$\quad\quad\quad\quad\quad\quad\quad\quad\quad*\mathbf{u1}$
$\quad + (q1*q2*c13*c21 - q1*q2*c11*c23$
$\quad\quad + q1*q3*c13*c31 - q1*q3*c11*c33)$
$\quad\quad\quad\quad\quad\quad\quad\quad\quad*\mathbf{u2}$
$\quad + (q1*q2*c11*c22 - q1*q2*c12*c21$
$\quad\quad + q1*q3*c11*c32 - q1*q3*c12*c31)$
$\quad\quad\quad\quad\quad\quad\quad\quad\quad*\mathbf{u3}$

On the other hand
v2+v3 = q2*(c21***u1** +c22***u2** +c23***u3**
    + q3*(c31***u1** +c32***u2** +c33***u3**)
  = (q2*c21+ q3*c31)***u1**
     +(q2*c22+q3*c32)***u2**
     +(q2*c23+q2*c33)***u3**

So
v1∧(v2+v3)
  = **v1**∧(q2*c21+ q3*c31)***u1**
         + q2*c22+ q3*c32)***u2**
         + q2*c23+ q3*c33)***u3**
  = (q1*c11***u1** + q1*c12***u2** +q1*c13***u3**)
     ∧ (q2*c21+ q3*c31)***u1**
         + q2*c22+ q3*c32)***u2**
         + q2*c23+ q3*c33)***u3**)
  = (q1*c11***u1** +q1*c12***u2** +q1*c13***u3**)
     ∧ (q2*c21+ q3*c31)***u1**
         + q2*c22+ q3*c32)***u2**
         + q2*c23+ q3*c33)***u3**)
  = (q1*c12*(q2*c23+ q3*c33)
         -q1*c13*(q2*c22+ q3*c32))***u1**
    +(q1*c13*(q2*c21+ q3*c31)
         -q1*c11*(q2*c23+ q3*c33))***u2**
    +(q1*c11*(q2*c22+ q3*c32)
         -q1*c12*(q2*c21+ q3*c31))***u3**
  = (q1*c12*q2*c23+ q1*c12*q3*c33
    -q1*c13*q2*c22-q1*c13*q3*c32)
                              ***u1**
    +(q1*c13*q2*c21+ q1*c13*q3*c31
     -q1*c11*q2*c23-q1*c11*q3*c33)
                              ***u2**
    +(q1*c11*q2*c22+q1*c11*q3*c32
     -q1*c12*q2*c21-q1*c12*q3*c31)
                              ***u3**

So they are equal."

Einstein, capitulating indirectly: " Make a table to show the individual summands so I can follow the equality more clearly.

To this demand Newton quickly produced the following table

$(v1 \wedge v2) + (v1 \wedge v3)$	$v1 \wedge (v2+v3)$
u1	
+q1*q2*c12*c23 − q1*q2*c13*c22 +q1*q3*c12*c33 − q1*q3*c13*c32	+ q1*c12*q2*c23 + q1*c12*q3*c33 −q1*c13*q2*c22 −q1*c13*q3*c32
u2	
+(q1*q2*c13*c21 − q1*q2*c11*c23 + q1*q3*c13*c31 − q1*q3*c11*c33	+q1*c13*q2*c21 + q1*c13*q3*c31 −q1*c11*q2*c23 −q1*c11*q3*c33
u3	
+(q1*q2*c11*c22 − q1*q2*c12*c21 + q1*q3*c11*c32 − q1*q3*c12*c31	+(q1*c11*q2*c22 +q1*c11*q3*c32 −q1*c12*q2*c21 −q1*c12*q3*c31

Einstein: "Now I can see the equality, although the multiplicative factors for the summands have been rearranged."

Breton: "We have here an example of an invalid proof contrasted with a valid proof."

Sums of outer products

Einstein, taking the lead again: "What can we say about
$$v3 \cdot [v1 * v2] + v3 \cdot [v4 * v5]?$$

Newton: "That's easy.
$$v3 \cdot [v1 * v2] = (v3 \cdot v1) * v2$$
and
$$v3 \cdot [v4 * v5] = (v3 \cdot v4) * v5$$
so
$$v3 \cdot [v1 * v2] + v3 \cdot [v4 * v5]$$
$$= (v3 \cdot v1) * v2 + (v3 \cdot v4) * v5$$

Breton: "So the outer multiplication is not so mysterious! Can these outer transformations be added?"

Einstein: "If so, then suppose
$$[v1 * v2] + [v4 * v5] = [(v1+v4) * (v2+v5)].$$

Newton: "Then
$$\mathbf{v3} \cdot ([\mathbf{v1}*\mathbf{v2}] + [\mathbf{v4}*\mathbf{v5}])$$
$$= \mathbf{v3} \cdot [\mathbf{v1}*\mathbf{v2}] + \mathbf{v3} \cdot [\mathbf{v4}*\mathbf{v5}]$$
$$= (\mathbf{v3} \cdot \mathbf{v1})*\mathbf{v2} + (\mathbf{v3} \cdot \mathbf{v4})*\mathbf{v5}]$$
while
$$\mathbf{v3} \cdot [(\mathbf{v1}+\mathbf{v4})*(\mathbf{v2}+\mathbf{v5})]$$
$$= \mathbf{v3} \cdot (\mathbf{v1}+\mathbf{v4})*(\mathbf{v2}+\mathbf{v5})$$
$$= (\mathbf{v3} \cdot \mathbf{v1}+\mathbf{v3} \cdot \mathbf{v4})*(\mathbf{v2}+\mathbf{v5})$$

$$= (\mathbf{v3} \cdot \mathbf{v1}+\mathbf{v3} \cdot \mathbf{v4})*\mathbf{v2}$$
$$+ (\mathbf{v3} \cdot \mathbf{v1}+\mathbf{v3} \cdot \mathbf{v4})*\mathbf{v5}$$
so it does *not* appear that outer products can be added."

Breton: "Not as outer products, but perhaps the sum of two outer products can result in another kind of transformation."

The vector triple product

Breton: "How about $\mathbf{v1} \wedge (\mathbf{v2} \wedge \mathbf{v3})$? The cross product $\mathbf{v2} \wedge \mathbf{v3}$ is itself a vector and as such can form a multiplicand with a third vector. So it is a legitimate addition to Newton's table. But what does it equal?"

Newton: "It's not obvious to me."

Breton: "Nor to me. Let me try to analyze the question. We know
$$(\mathbf{v2} \wedge \mathbf{v3}) = qv2*qv3*\sin(\text{angle}(2,3))*\mathbf{un}(2,3)$$
$$\mathbf{v1} \wedge (\mathbf{v2} \wedge \mathbf{v3}) = qv1*qv2*qv3*\sin(\text{angle}(2,3))*\mathbf{uv1} \wedge \mathbf{un}(2,3)$$
Remember $\mathbf{un}(2,3)$ is orthogonal to both $\mathbf{v2}$ and $\mathbf{v3}$. That is,
$$\mathbf{uv2} \cdot \mathbf{un}(2,3) = 0 = \mathbf{uv3} \cdot \mathbf{un}(2,3)$$
Furthermore
$$\mathbf{uv1} \cdot (\mathbf{uv1} \wedge \mathbf{un}(2,3)) = \mathbf{un}(2,3) \cdot (\mathbf{uv1} \wedge \mathbf{uv1})$$
$$= 0$$
so the vector $\mathbf{uv1} \wedge \mathbf{un}(2,3)$ is orthogonal to $\mathbf{uv1}$. So also to $\mathbf{un}(2,3)$. Therefore $\mathbf{v1} \wedge \mathbf{un}(2,3)$ must lie in the plane generated by $\mathbf{v2}$ and $\mathbf{v3}$. So I conclude that
$$\mathbf{v1} \wedge (\mathbf{v2} \wedge \mathbf{v3}) = a*\mathbf{v2} + b*\mathbf{v3}$$
for some scalars *a* and *b*."

Newton: "Not bad. So we need only determine two scalar quantities, *a* and *b*."

Einstein: "The factor, $qv1*qv2*qv3$, shows we are dealing with some kind of volume which is not a scalar like $\mathbf{v1} \cdot (\mathbf{v2} \wedge \mathbf{v3})$ but a vector, a vector-volume. Is this the same as a volume of vectors?"

Breton: "You have an inquisitive mind Einstein. Like any vector a vectorial volume has a magnitude and a direction. We have discovered that the vectorial volume **v1**∧(**v2**∧**v3**) can be decomposed into two other vectorial volumes, one in the direction **uv2** and another in the direction **uv3**. But let us put your question aside for now as a distraction. Right now we are trying to obtain an equation for Newton's table."

Newton: "Which we have reduced to determining two scalars, *a* and *b*. It seems to me that both *a* and *b* must somehow involve **v1**."

Breton: "You have good instincts, Newton. Furthermore, *a* and *b* must both be scalar 'areas'."

Newton: "How do we proceed?"

Breton: "Let's start with some examples which may show us the way. The path ahead looks rough, hard to cut through. And let's just consider the directions because we know the factor qv1∗qv2∗qv3 will finally apply to the full volume. So first look at a cube where the edges and their directions coincide the vectors. Here look at this diagram.

With that Breton handed his friends this sketch.

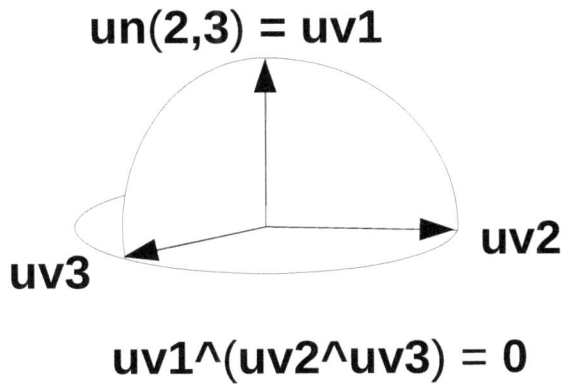

Breton: "You are looking at a corner of the cube whether from the inside of from the outside makes no difference. For this geometry **v2** and **v3** are orthogonal to each other and **v1** is orthogonal to both. For this example,
$$(uv2 \wedge uv3) = uv1$$
since sin(angle(2,3)) = 1.
Then
$$uv1 \wedge (uv2 \wedge uv3) = uv1 \wedge uv1 = 0$$

85

Newton: "Not much learned from this example. The result would hold for any rectangular parallelepiped. Another example?"

Breton: "Let's incline **uv1** in the **uv3** direction Then again(**uv2**∧**uv3**) = **un**(2,3) as in this next sketch.

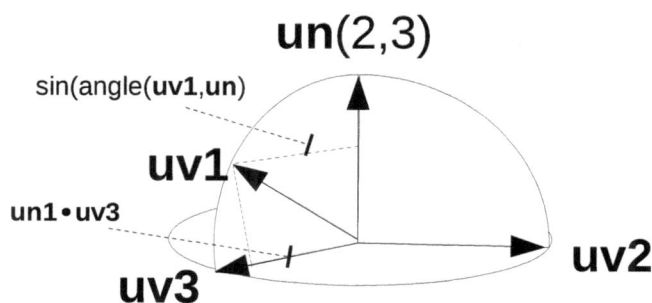

$$\mathbf{uv1} \wedge (\mathbf{uv2} \wedge \mathbf{uv3}) = (\mathbf{uv1} \cdot \mathbf{uv3}) * \mathbf{uv2}$$

Now
$$\mathbf{uv1} \wedge (\mathbf{un}(2,3)) = \sin(\text{angle}(1,\text{un})) * \mathbf{uv2}.$$
Notice sin(angle(1,un)) = cos(angle(1,3))."

Newton: "And cos(angle(1,3)) = **uv1**•**uv3**. So for this example
$$\mathbf{uv1} \wedge (\mathbf{uv2} \wedge \mathbf{uv3}) = (\mathbf{uv1} \cdot \mathbf{uv3}) * \mathbf{uv2}.$$
Again this example could be expanded to parallelepipeds similarly inclined."

Breton: "We might have inclined **uv1** in the **uv2** direction and so obtained
$$\mathbf{uv1} \wedge (\mathbf{uv2} \wedge \mathbf{uv3}) = (\mathbf{uv1} \cdot \mathbf{uv2}) * \mathbf{uv3}.$$

Einstein: "Wait a minute. The cross product could be positive or negative. Which one is it here?"

Breton: "Very little gets by you Einstein. The two products,
(**uv1**•**uv3**)∗**uv2** and (**uv1**•**uv2**)∗**uv3**
might both be positive, or both negative, or one positive and the other negative."

Newton: "I suspect we can sharpen the result a little. Recall we found previously that the cross products could be organized into two groups by sign. Within each group the products kept a cyclic rotation. Here we see the products do not keep the same cyclic rotation, so I suspect the products differ in sign. Furthermore (**uv1**•**uv2**)∗**uv3** will be positive while (**uv1**•**uv3**)∗**uv2** will be negative."

Breton: "For now let's continue our search for a comprehensive solution. Suppose **v2** and **v3** are not orthogonal, while **v1** is orthogonal to both.

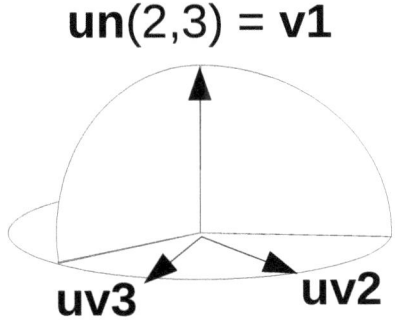

$$uv1 \wedge (uv2 \wedge uv3) = 0$$

Eintein: "Not much different from the fully orthogonal case since **uv2**∧**uv3** = sin(angle(2,3))*__uv1__ and **uv1**∧**uv1** = **0**."

Breton: "Good! And if **uv1** is inclined toward **uv2** or **uv3** we get the same answer as as before. So of all the possible case we might explore, only one is left—the case where **uv1** is inclined arbitrarily."

With that Breton handed the following illustration to his friends.

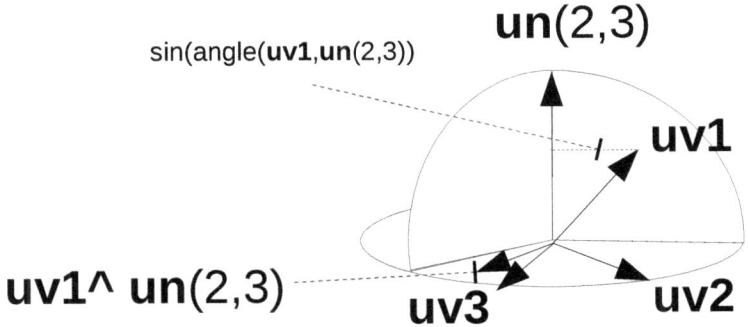

Breton: "The vector **uv1**∧**un**(2,3) is orthogonal to both **uv1** and **un**."

Einstein: "But **uv1**∧**un**(2,3) does not equal **uv1**∧(**uv2**∧**uv3**)!"

Breton: "True enough, but close. As we have seen
$$uv1 \wedge (uv2 \wedge uv3) = \sin(\text{angle}(2,3)) * uv1 \wedge un(2,3)$$
so we seek a vector parallel to the one illustrated."

Einstein: "In fact,
$$uv1 \wedge (uv2 \wedge uv3)$$
$$= \sin(\text{angle}(2,3)) * \sin(\text{angle}(uv1, un(2,3)))$$
$$* un(uv1, un(2,3))$$

Newton: "So we have the right direction and magnitude. What more do we want?"

Breton: "You have noted the result is a vector in the plane defined by **uv2** and **uv3**. Can we express the result as a vector in that plane?"

Einstein: "Breton draw a diagram showing all the vectors!"

Breton: "All right, but it could be complicated. Within a few minutes Breton presented his friends the following sketch."

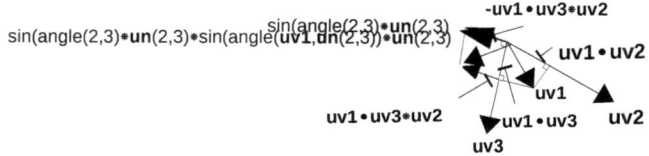

Einstein: "What's this? The illustration is way too complicated. I can't understand it."

Breton: "As I suspected. A geometric proof appears to be quite complicated."

Newton: " Then let's try a proof with an assumed orientation."

Einstein: "Let me try this time. First assume an orientation specifying **u1**, **u2**, and **u3**. Express
$$v1 \text{ as } q1*(c11*u1 + c12*u2 + c13*u3$$
$$v2 \text{ as } q2*(c21*u1 + c22*u2 + c23*u3$$
$$v3 \text{ as } q3*(c31*u1 + c32*u2 + c33*u3$$
Then
$$v2 \wedge v3 = q2*q3*((c22*c33 - c23*c32)*u1$$
$$+ (c23*c31 - c21*c33)*u2$$
$$+ (c21*c32 - c22*c31)*u3)$$

So
v1∧(**v2**∧**v3**)
 = (q1*(c11***u1** +c12***u2** +c13***u3**)
   ∧ (q2*q3*((c22*c33 − c23*c32)***u1**
            + (c23*c31 − c21*c33)***u2**
            + (c21*c32 − c22*c31)***u3**))

Breton, this is complicated."

Breton: "Just continue the bookkeeping until you get a result expressed in the orientation."

Einstein: "Then check that I make no arithmetic errors.
v1∧(**v2**∧**v3**)
 = (q1*q2*q3*(c11***u1** +c12***u2** +c13***u3**)
      ∧ ((c22*c33 − c23*c32)***u1**
         + (c23*c31 − c21*c33)***u2**
         + (c21*c32 − c22*c31)***u3**))
 = (q1*q2*q3
    (c12(c21*c32 − c22*c31)
         − c13*(c23*c31 − c21*c33))***u1**
    + (c13*(c22*c33 − c23*c32)
         − c11*(c21*c32 − c22*c31))***u2**
    + (c11*(c23*c31 − c21*c33)
         − c12*(c22*c33 − c23*c32))***u3**)

So here is **v1**∧(**v2**∧**v3**) expressed in coordinates. But how does this relate to **uv2** and **uv3**?"

Breton: "Remember Newton's rhomboid? Our problem has been reduced to one where we know the sum of two vectors, each of which we know the direction, but not the magnitude."

With that Breton sketched the following diagram for his friends.

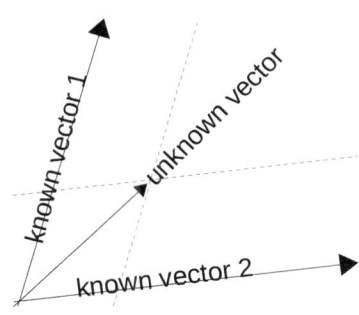

Breton: "See how the unknown magnitudes can become known by constructing parallel lines to the two vectors."

Einstein: "Show us how this solves our problem."

Breton: "Remember
$$\mathbf{v1} \wedge (\mathbf{v2} \wedge \mathbf{v3}) = a*\mathbf{v2} + b*\mathbf{v3}$$
for some scalars a and b."

Einstein: "We know the solution lies in the plane defined by **v1** and **v2**. But any two different vectors in the plane also define the same plane. So solutions can exist without any reference to **v1** and **v2**. For any two vectors in the plane say **x** and **y** we could find some a and b such that
$$\mathbf{v1} \wedge (\mathbf{v2} \wedge \mathbf{v3}) = a*\mathbf{x} + b*\mathbf{y}.$$
Moreover, if I fix a***x** then
$$b*\mathbf{y} = \mathbf{v1} \wedge (\mathbf{v2} \wedge \mathbf{v3}) - a*\mathbf{x}.$$

Breton: "So there are an infinite number of solutions to our problem. In fact even specifying **v2** and **v3**, which Newton suggests, there still exist and infinite number. However, if we choose a***x** then there exists only one solution for b***y**. So, Einstein, you have shown us a way to solve our problem."

Einstein is flabbergasted, so much so he refuses to continue.

So Newton takes up the discussion. Newton: "Let us choose a*$\mathbf{v2}$ as $\mathbf{v1} \bullet \mathbf{v3} * \mathbf{v2}$. What is b?"

Breton: "As Einstein stated,
$$b*\mathbf{v3} = \mathbf{v1} \wedge (\mathbf{v2} \wedge \mathbf{v3}) - \mathbf{v1} \bullet \mathbf{v3} * \mathbf{v2}.$$
So the value of b can be calculated."

Newton: "Let's do it. First express $\mathbf{v1} \bullet \mathbf{v3} * \mathbf{v2}$ in terms of the given orientation.
$\mathbf{v1} \bullet \mathbf{v3}$ = q1*(c11***u1** +c12***u2** +c13***u3**
    •q3*(c31***u1** +c32***u2** +c33***u3**)
  = q1*q3*(c11*c31 + c12*c32 + c13*c33)
so
$\mathbf{v1} \bullet \mathbf{v3} * \mathbf{v2}$= q1*q3*(c11*c31 + c12*c32 + c13*c33)
   q2(c21***u1** +c22***u2** +c23***u3**
Then
$\mathbf{v1} \wedge (\mathbf{v2} \wedge \mathbf{v3}) - \mathbf{v1} \bullet \mathbf{v3} * \mathbf{v2}$
  = (q1*q2*q3
   (c12(c21*c32 − c22*c31)
     − c13*(c23*c31 − c21*c33))***u1**
   + (c13*(c22*c33 − c23*c32)
     − c11*(c21*c32 − c22*c31))***u2**
   + (c11*(c23*c31 − c21*c33)
     − c12*(c22*c33 − c23*c32))***u3**)

$$- q_1*q_2*q_3*(c_{11}*c_{31} + c_{12}*c_{32} + c_{13}*c_{33})$$
$$*(c_{21}*\mathbf{u_1} + c_{22}*\mathbf{u_2} + c_{23}*\mathbf{u_3})$$

Einstein: "That's a lot of symbols."

Breton: "Just a lot of bookkeeping. Keep going until you can express the result in the orientation."

Newton: "Please check I don't make arithmetical errors.
v1∧(v2∧v3) − v1•v3*v2
$$= (q_1*q_2*q_3$$
$$*(c_{12}*(c_{21}*c_{32} - c_{22}*c_{31})$$
$$- c_{13}*(c_{23}*c_{31} - c_{21}*c_{33}))*\mathbf{u_1}$$
$$- (c_{11}*c_{31} + c_{12}*c_{32} + c_{13}*c_{33})*c_{21}*\mathbf{u_1}$$
$$+ (c_{13}*(c_{22}*c_{33} - c_{23}*c_{32})$$
$$- c_{11}*(c_{21}*c_{32} - c_{22}*c_{31}))*\mathbf{u_2}$$
$$- (c_{11}*c_{31} + c_{12}*c_{32} + c_{13}*c_{33})*c_{22}*\mathbf{u_2}$$
$$+ (c_{11}*(c_{23}*c_{31} - c_{21}*c_{33})$$
$$- c_{12}*(c_{22}*c_{33} - c_{23}*c_{32}))*\mathbf{u_3})$$
$$- (c_{11}*c_{31} + c_{12}*c_{32} + c_{13}*c_{33})*c_{23}*\mathbf{u_3}$$
$$= (q_1*q_2*q_3$$
$$*(c_{12}*c_{21}*c_{32} - c_{12}*c_{22}*c_{31}$$
$$- c_{13}*c_{23}*c_{31} + c_{13}*c_{21}*c_{33}$$
$$- c_{11}*c_{31}*c_{21} - c_{12}*c_{32}*c_{21}$$
$$- c_{13}*c_{33}*c_{21})$$
$$*\mathbf{u_1}$$
$$+ (c_{13}*c_{22}*c_{33} - c_{13}*c_{23}*c_{32}$$
$$- c_{11}*c_{21}*c_{32} + c_{11}*c_{22}*c_{31}$$
$$- c_{11}*c_{31}*c_{22} - c_{12}*c_{32}*c_{22}$$
$$- c_{13}*c_{33}*c_{22})$$
$$*\mathbf{u_2}$$
$$+ (c_{11}*c_{23}*c_{31} - c_{11}*c_{21}*c_{33}$$
$$- c_{12}*c_{22}*c_{33} + c_{12}*c_{23}*c_{32}$$
$$- c_{11}*c_{31}*c_{23} - c_{12}*c_{32}*c_{23}$$
$$- c_{13}*c_{33}*c_{23})$$
$$*\mathbf{u_3}$$

What can we make of this mess?"

Breton: "I see some terms can be canceled. For instance in the **u1** direction
$$c_{12}*c_{21}*c_{32} - c_{12}*c_{32}*c_{21} = 0.$$
There are more. Continue by taking out the canceled terms."

Newton: "Alright.
v1∧(v2∧v3) − v1•v3*v2
$$= (q_1*q_2*q_3$$
$$*((- c_{12}*c_{22}*c_{31}$$
$$- c_{13}*c_{23}*c_{31}$$
$$- c_{11}*c_{31}*c_{21})*\mathbf{u_1}$$

$$+ (- c13*c23*c32$$
$$- c11*c21*c32$$
$$- c12*c32*c22)*\mathbf{u2}$$
$$+ (- c11*c21*c33$$
$$- c12*c22*c33$$
$$- c13*c33*c23)*\mathbf{u3})$$
$$= - (q1*q2*q3$$
$$*((c12*c22 + c13*c23 + c11*c21)*c31*\mathbf{u1}$$
$$+ (c12*c22 + c13*c23 + c11*c21)*c32*\mathbf{u2}$$
$$+ (c11*c21 + c12*c22 + c13*c23)*c33*\mathbf{u3})$$

So we get fewer symbols, so what?"

Breton: "Don't you see that
$$c12*c22 + c13*c23 + c11*c21$$
$$c12*c22 + c13*c23 + c11*c21$$
$$c11*c21 + c12*c22 + c13*c23$$
have the same value? They can be factored.

Newton: "Then we would get
$$\mathbf{v1} \wedge (\mathbf{v2} \wedge \mathbf{v3}) - \mathbf{v1} \bullet \mathbf{v3} * \mathbf{v2}$$
$$= - (q1*q2*q3)$$
$$*(c11*c21 + c12*c22 + c13*c23)$$
$$*(c31*\mathbf{u1} + c32*\mathbf{u2} + c33*\mathbf{u3})$$
And what is this?"

Breton: "And what is $\mathbf{v1} \bullet \mathbf{v2}$ and $\mathbf{v3}$?"

Newton: "I see
$$\mathbf{v1} \wedge (\mathbf{v2} \wedge \mathbf{v3}) - \mathbf{v1} \bullet \mathbf{v3} * \mathbf{v2}$$
$$= - (q1*q2$$
$$*(c11*c21 + c12*c22 + c13*c23)$$
$$*q3*(c31*\mathbf{u1} + c32*\mathbf{u2} + c33*\mathbf{u3})$$
$$= -\mathbf{v1} \bullet \mathbf{v2} * \mathbf{v3}$$

Breton: "So you can conclude
$$\mathbf{v1} \wedge (\mathbf{v2} \wedge \mathbf{v3}) = \mathbf{v1} \bullet \mathbf{v3} * \mathbf{v2} - \mathbf{v1} \bullet \mathbf{v2} * \mathbf{v3}!$$

Einstein, resuscitated: "I like the result. It has a certain symmetry which is easy to remember."

Breton: "Remember that
$$\mathbf{v1} \bullet (\mathbf{v2} + \mathbf{v3}) = (\mathbf{v1} \bullet \mathbf{v2}) + (\mathbf{v1} \bullet \mathbf{v3})$$
could be interpreted geometrically as two separate vectorial *areas* equaling a third. Newton has just shown two separate vectorial *volumes* equaling a third vectorial volume.

Newton: "Something similar could be said of our other results. These results would be very, very difficult to prove geometrically."

Breton: "Referencing the origin requires attention to details, but in the end results in a much simpler proof."

Newton: "I see something interesting. Notice
$$v1 \bullet v3*v2 = v1 \bullet [v3*v2]$$
an outer product. So
$$v1 \wedge (v2 \wedge v3) = v1 \bullet [v3*v2 - v2*v3]$$
which involves the difference of two outer products."

Breton: "Brilliant. Your observation should prove useful in the future."

Einstein, in reluctant resignation: "Yes, but Breton haven't you assumed Euclidean plane geometry? So does our vector algebra rest on Euclid's axioms?"

Newton: "What an amazingly wonderful tribute to my illustrious ancestor."

Breton: "Not so. It is a point worth discussing. For his geometry Euclid includes an axiom called the **parallel postulate**, namely that parallel lines never meet. From this postulate he easily deduces that the interior angles of a triangle equal two right angles. This is the language of Euclidean geometry, a mathematical science.

The language of vectorial calculus is somewhat different. Parallel lines for this admittedly mathematical science are those which have the same direction. So even though the two sciences use the same word, *parallel*, the meaning of the word differs in each. So it is with many of the words we have been using like lines, points, etc. which are axioms for Euclid, but in vectorial calculus are derivative ideas based on direction, magnitude, and the underlying field with its own calculus and topology.

So our words have been confusing two different dictionaries, somewhat like the confusion between the Mathematics and Theoretical Physics."

Newton: "Both dictionaries, the one for Euclidean geometry and the one for vectorial algebra are mathematical dictionaries."

Breton: "For this reason we are tempted to use the same words for different ideas. We saw in tp1.1 that any science cannot tolerate the ambiguity."

Newton: "They are specialized sub-dictionaries. Still, we can define ideas for vectorial algebra which coincide with the axioms of Euclidean Geometry."

Breton: "Yes, we can. The same ideas remain axioms in Euclidean Geometry, but are derivative ideas in vectorial algebra."

Einstein: "It's a subtle point indeed."

Breton: "Again we see the difference between words and ideas. One can conceive right angles even in non-planar geometries. Moreover, even in a plane one can conceive geometries which are non-Euclidean. Although, when we *defined* triangles, we used the fact that in the Euclidean plane the sum of the angles in a triangle equals 180 degrees. These special definitions referred only to Euclidean plane triangles. Angles and triangles may be defined in a broader context. In this broader context, our results will still hold because the logic remains the same for these different triangles. In effect we used Euclidean Geometry to show specific results that apply generically. In the generic case, the axiomatic ideas of Euclidean Geometry would have to be defined in terms of vectorial algebra: direction, magnitude, and the scalar field. That said, Euclidean geometry does comport well with vector algebra."

Newton: "Previously we saw that vectorial proofs can facilitate geometrical proofs. Sometimes we see the reverse—geometrical proofs facilitate a vectorial conclusion. Vectorial algebra and Geometry march together like a groom and bride."

Breton: "Newton, would you update your table to includes our latest results?"

Axiomatic	Comments
$v1+v2 = v3$	closure
$q*v1 = v2$	Scalar multiply
$v1+(v2+v3) = (v1+v2)+v3$	association
Defined: two at a time	
$v1 \cdot v2 = v2 \cdot v1$	Inner product
$b*v1 \cdot c*v2 = b*c*(v1 \cdot v2)$	
$abs(v1 \cdot v2) \leq abs(v1) * abs(v2)$	
$v1 \wedge v2 = -(v2 \wedge v1)$ $= ((-v2) \wedge v1)$ $= (v2 \wedge (-v1))$	cross product
$v1 \wedge v1 = 0$	
$(b*v1) \wedge (c*v2) = b*c*(v1 \wedge v2)$	
$abs(v1 \wedge v2) \leq abs(v1)*abs(v2)$	
$v1 \cdot (v1 \wedge v2) = v2 \cdot (v1 \wedge v2) = 0$	
$(b*v1)*(c*v2) = b*c*(v1*v2)$	
$(abs(v1)*abs(v2))^2$ $= (abs(v1 \wedge v2))^2$ $+ (abs(v1 \cdot v2))^2$	
Defined: three at at time	
$v1 \cdot (v2+v3) = v1 \cdot v2 + v1 \cdot v3$	
$v1 \wedge (v2+v3) = v1 \wedge v2 + v1 \wedge v3$	
$v1*(v2+v3) = v1*v2 + v1*v3$	
$v1 \cdot (v2 \wedge v3) = v2 \cdot (v3 \wedge v1)$ $= v3 \cdot (v1 \wedge v2)$ $= (v1 \wedge v2) \cdot v3$	Scalar triple product

$= (v2 \wedge v3) \cdot v1$ $= (v3 \wedge v1) \cdot v2$	
$v1 \cdot (v2 * v3) = (v1 \cdot v2) * v3$	transformation
$v1 \wedge (v2 \wedge v3) = (v1 \cdot v3) * v2$ $\quad - (v1 \cdot v2) * v3$ $= v3 \cdot (v1 * v2)$ $\quad - v3 * (v1 \cdot v2)$ $= v1 \cdot (v3 * v2$ $\quad - v1 \cdot (v2 * v3)$	Vector triple pr duct

Einstein: "I see you have added some other results."

Newton: "Yes. They are simply elaborations from earlier results. For instance
$$(v1 \cdot v3) * v2 = v3 \cdot (v1 * v2)$$
since
$$(v1 \cdot v3) = v3 \cdot v1$$
and
$$(v3 \cdot v1) * v2 = v3 \cdot (v1 * v2).$$

Einstein: "You should add more because $v2 \wedge v3 = -v3 \wedge v2$. So we can put
$$v1 \wedge (v3 \wedge v2) = (v1 \cdot v2) * v3 - (v1 \cdot v3) * v2.$$

Newton: "Of course. Now I can see several other valid combinations as well. Let me update the table to include them too."

After a few minutes Newton produced the additions to his table.

Defined: three at at time	
$(v1 \wedge v2) \wedge v3$ $\quad = (v1 \cdot v3) * v2 - (v2 \cdot v3) * v1$ $\quad = v1 \cdot (v3 * v2)$ $\qquad - v1 * (v3 \cdot v2)$ $\quad = v3 \cdot (v1 * v2 - v2 * v1)$	
$v1 \wedge (v2 \wedge v3) - (v1 \wedge v2) \wedge v3$ $\quad = v2 \cdot (v3 * v1 - v1 * v3)$ $v1 \wedge (v2 \wedge v3)$ $\quad + v2 \wedge (v3 \wedge v1)$ $\quad + v3 \wedge (v1 \wedge v2) = 0$	

Newton: "The top entry uses your remark, Einstein, and relabels some of the vectors. The second entry finds an identity in combinations of the differences. Now
$$v1 \wedge (v2 \wedge v3) = (v1 \cdot v3) * v2 - (v1 \cdot v2) * v3$$
$$(v1 \wedge v2) \wedge v3 = (v1 \cdot v3) * v2 - (v2 \cdot v3) * v1$$

so their difference
$$v1 \wedge (v2 \wedge v3) - (v1 \wedge v2) \wedge v3 = -(v1 \cdot v2)*v3 + (v2 \cdot v3)*v1$$
$$= v2 \cdot (v3*v1 - v1*v3)$$
The third item uses relabeling to discover a remarkable identity.
$$v1 \wedge (v2 \wedge v3) = (v1 \cdot v3)*v2 - (v1 \cdot v2)*v3$$
$$v2 \wedge (v3 \wedge v1) = (v2 \cdot v1)*v3 - (v2 \cdot v3)*v1$$
$$v3 \wedge (v1 \wedge v2) = (v3 \cdot v2)*v1 - (v3 \cdot v1)*v2$$
so their vectorial sum equals the zero vector."

Breton: "Your table shows we are truly on the way to developing a vectorial algebra."

Einstein: "We still have some way to go. Let's examine vectors four at a time."

Newton: "Some of these are straightforward. Try
$$(v1+v2) \cdot (v3+v4)$$
$$= v1 \cdot (v3+v4) + v2 \cdot (v3+v4)$$
$$= v1 \cdot v3 + v1 \cdot v4 + v2 \cdot v3 + v2 \cdot v4$$

Einstein: "Straightforward enough."

Newton: "Then how about
$$(v1+v2) \wedge (v3+v4)$$
$$= v1 \wedge v3 + v1 \wedge v4 + v2 \wedge v3 + v2 \wedge v4$$
and
$$(v1+v2)*(v3+v4)$$
$$= v1*v3 + v1*v4 + v2*v3 + v2*v4$$

Einstein: "Yes, just use the same argument."

Newton: "Let's try something more difficult. Consider the scalar
$$(v1 \wedge v2) \cdot (v3 \wedge v4)$$

Einstein: "Not so difficult.
$$(v1 \wedge v2) \cdot (v3 \wedge v4) = q1*q2*\sin(\text{angle}1,2)$$
$$*q3*q4*\sin(\text{angle}3,4)$$
$$*\mathbf{un}(1,2) \cdot \mathbf{un}(3,4)$$
$$= q1*q2*q3*q4$$
$$*\sin(\text{angle}1,2)*\sin(\text{angle}3,4)$$
$$*\cos(\mathbf{un}(1,2),\mathbf{un}(3,4)).$$

Breton: "Again we can ask for this relationship in terms of a purely vector equation, but our experience with triple products forewarns us of no small difficulties. May I suggest we abandon this part of the trail for now to take it up later when we have ascended a little further."

Einstein: "Again promises, promises."

Breton: "Which will be kept anon."

Einstein: "What good are these expansions? They seems so much more clumsy.

Breton: "It often makes proving propositions rather more simple. Let me illustrate. We have shown for the triple scalar product.
$$(v1 \wedge v2) \cdot v3 = (v2 \wedge v3) \cdot v1$$
basing the proof on geometry. Let's try a proof based on the origin.

We would have
$(v1 \wedge v2) \cdot v3$
 = q1*q2*((c12*c23 − c13*c22)***u1**)
    + (c13*c21 − c11*c23)***u2**
    + (c11*c22 − c12*c21)***u3**)
  •q3*(c31***u1** + c32***u2** + c33***u3**)
where **v3** = q3*(c31***u1** + c32***u2** + c33***u3**)
Then
$(v1 \wedge v2) \cdot v3$
 = q1*q2*q3
  *((c12*c23 − c13*c22)*c31
   + (c13*c21 − c11*c23)*c32
   + (c11*c22 − c12*c21)*c33)
and
$(v2 \wedge v3) \cdot v1$
 = q2*q3*((c22*c33 − c23*c32)***u1**)
    + (c23*c31 − c21*c33)***u2**
    + (c21*c32 − c22*c31)***u3**)
  •q1*(c11***u1** + c12***u2** + c13***u3**)
 = q2*q3*q1
  *(c22*c33 − c23*c32)*c11
   + (c23*c31 − c21*c33)*c12
   + (c21*c32 − c22*c31)*c13)
Are they equal?"

Newton: "They are not the same, but they could be equal. Let me assemble the factors.
For $(v1 \wedge v2) \cdot v3$
(c12*c23 − c13*c22)*c31
 + (c13*c21 − c11*c23)*c32
 + (c11*c22 − c12*c21)*c33)
  = c12*c23*c31
   +c13*c21*c32
   +c11*c22*c33
    − c13*c22*c31
    − c11*c23*c32
    − c12*c21*c33)

For $(\mathbf{v2} \wedge \mathbf{v3}) \cdot \mathbf{v1}$
$(c22*c33 - c23*c32)*c11$
$\quad + (c23*c31 - c21*c33)*c12$
$\quad + (c21*c32 - c22*c31)*c13)$
$\quad\quad = c22*c33*c11$
$\quad\quad\quad + c23*c31*c12$
$\quad\quad\quad + c21*c32*c13$
$\quad\quad\quad\quad - c23*c32*c11$
$\quad\quad\quad\quad - c21*c33*c12$
$\quad\quad\quad\quad - c22*c31*c13$

So although the summands are differently ordered, each summand in one case finds its match in the other case.
And in both cases
$$q1*q2*q3 = q2*q3*q1$$
So $(\mathbf{v2} \wedge \mathbf{v3}) \cdot \mathbf{v1}$ does indeed equal $(\mathbf{v1} \wedge \mathbf{v2}) \cdot \mathbf{v3}$."

Einstein, concluding: "A proof without the difficult geometry."

Breton: "The origin with its orthogonal coordinates does not eliminate the geometry so much as finesses it. True geometrical propositions about areas and volumes, difficult to prove by geometrical methods alone, may find easier proofs using vectorial algebra.

Einstein: "I am willing to concede that all the variations of the scalar triple product can be proven similarly, but what about the vector triple product?"

Newton: "Let me try
$\mathbf{v1} \wedge (\mathbf{v2} \wedge \mathbf{v3}) = q1*q2*q3 * \mathbf{uv1} \wedge (\mathbf{uv2} \wedge \mathbf{uv3})$
and
$(\mathbf{uv1} \cdot \mathbf{uv3})*\mathbf{uv2} - (\mathbf{uv1} \cdot \mathbf{uv2})*\mathbf{uv3}$
$\quad = q1*q3*q2*(\mathbf{uv1} \cdot \mathbf{uv3})*\mathbf{uv2}$
$\quad\quad - q1*q2*q3(\mathbf{uv1} \cdot \mathbf{uv2})*\mathbf{uv3}$
so the factor $q1*q2*q3$ occurs in both expressions; we may then deal only with the directions.
$\mathbf{uv1} \wedge (\mathbf{uv2} \wedge \mathbf{uv3}) = (c11*\mathbf{u1} + c12*\mathbf{u2} + c13*\mathbf{u3})$
$\quad\quad \wedge ((c21*\mathbf{u1} + c22*\mathbf{u2} + c23*\mathbf{u3})$
$\quad\quad\quad \wedge (c31*\mathbf{u1} + c32*\mathbf{u2} + c33*\mathbf{u3}))$
$\quad = (c11*\mathbf{u1} + c12*\mathbf{u2} + c13*\mathbf{u3})$
$\quad\quad \wedge (c21*\mathbf{u1} \wedge c31*\mathbf{u1}$
$\quad\quad\quad + c22*\mathbf{u2} \wedge c31*\mathbf{u1}$
$\quad\quad\quad + c23*\mathbf{u3} \wedge c31*\mathbf{u1}$
$\quad\quad\quad + c21*\mathbf{u1} \wedge c32*\mathbf{u2}$
$\quad\quad\quad + c22*\mathbf{u2} \wedge c32*\mathbf{u2}$
$\quad\quad\quad + c23*\mathbf{u3} \wedge c32*\mathbf{u2}$
$\quad\quad\quad + c21*\mathbf{u1} \wedge c33*\mathbf{u3}$
$\quad\quad\quad + c22*\mathbf{u2} \wedge c\ c33*\mathbf{u3}$
$\quad\quad\quad + c23*\mathbf{u3} \wedge c33*\mathbf{u3})$

$$= (c11*u1 + c12*u2 + c13*u3)$$
$$\wedge(-c22c31*u3$$
$$+ c23*c31*u2$$
$$+ c21*c32*u3$$
$$- c23*c32*u1$$
$$- c21*c33*u2$$
$$+ c22*c33*u1)$$
$$= (c11*u1 + c12*u2 + c13*u3)$$
$$\wedge((c22*c33 - c23*c32)*u1$$
$$+ (c23*c31 - c21*c33)*u2$$
$$+ (c21*c32 - c22*c31)*u3)$$
$$= (c11*u1 \wedge (c22*c33 - c23*c32)*u1$$
$$+ c12*u2 \wedge (c22*c33 - c23*c32)*u1$$
$$+ c13*u3) \wedge (c22*c33 - c23*c32)*u1$$
$$+ c11*u1 \wedge (c23*c31 - c21*c33)*u2$$
$$+ c12*u2 \wedge (c23*c31 - c21*c33)*u2$$
$$+ c13*u3) \wedge (c23*c31 - c21*c33)*u2$$
$$+ c11*u1 \wedge (c21*c32 - c22*c31)*u3$$
$$+ c12*u2 \wedge (c21*c32 - c22*c31)*u3$$
$$+ c13*u3) \wedge (c21*c32 - c22*c31)*u3$$
$$= (c12*(c21*c32 - c22*c31)$$
$$- c13*(c23*c31 - c21*c33))*u1$$
$$+ (c13*(c22*c33 - c23*c32)$$
$$- c11*(c21*c32 - c22*c31))*u2$$
$$+ (c11*(c23*c31 - c21*c33)$$
$$- c12*(c22*c33 - c23*c32))*u3$$

So after careful bookkeeping we conclude
uv1∧(**uv2**∧**uv3**)
$$= (c12*(c21*c32 - c22*c31)$$
$$- c13*(c23*c31 - c21*c33))*u1$$
$$+ (c13*(c22*c33 - c23*c32)$$
$$- c11*(c21*c32 - c22*c31))*u2$$
$$+ (c11*(c23*c31 - c21*c33)$$
$$- c12*(c22*c33 - c23*c32))$$

Now let me calculate
(**uv1**•**uv3**)***uv2** - (**uv1**•**uv2**)***uv3**
$$= (c11*u1 + c12*u2 + c13*u3)$$
$$\bullet (c31*u1 + c32*u2 + c33*u3)$$
$$*(c21*u1 + c22*u2 + c23)*u3)$$
$$-(c11*u1 + c12*u2 + c13*u3)$$
$$\bullet (c21*u1 + c22*u2 + c23*u2)$$
$$*(c31*u1 + c32*u2 + c33)*u3$$
$$= (c11*c31 + c12*c32 + c13*c33)$$
$$*(c21*u1 + c22*u2 + c23)*u3)$$
$$-(c11*c21 + c12*c22 + c13*c23)\ c21*u1$$
$$*(c31*u1 + c32*u2 + c33)*u3$$

$$= ((c11*c31 + c12*c32 + c13*c33)*c21$$
$$-(c11*c21 + c12*c22 + c13*c23)*c31)*\mathbf{u1}$$
$$+((c11*c31 + c12*c32 + c13*c33)*c22$$
$$-(c11*c21 + c12*c22 + c13*c23)*c32)*\mathbf{u2}$$
$$+((c11*c31 + c12*c32 + c13*c33)*c23$$
$$-(c11*c21 + c12*c22 + c13*c23)*c33)*\mathbf{u3}$$
$$= (c12*c32*c21 - c12*c22*c31$$
$$+ c13*c33*c21 - c13*c23*c31)*\mathbf{u1}$$
$$+ ((c11*c31*c22 - c11*c21*c32$$
$$+ c13*c33*c22 - c13*c23*c32)*\mathbf{u2}$$
$$+ (c11*c31*c23 - c11*c21*c33$$
$$+ c12*c32*c23 - c12*c22*c33)*\mathbf{u3}$$

The two derivations match! So here is a different proof that
$$\mathbf{v1} \wedge (\mathbf{v2} \wedge \mathbf{v3}) = (\mathbf{uv1} \cdot \mathbf{uv3})*\mathbf{uv2} - (\mathbf{uv1} \cdot \mathbf{uv2})*\mathbf{uv3}.$$

Einstein: "So we have traded geometry for bookkeeping."

Breton: "Two different trails to the same place. We have climbed a little higher."

Einstein: "The bookkeeping trail is easier. Why not try it now on four vectors at a time?"

Other Identities

Newton: "Let's pick up where we were before. Does
$(\mathbf{v1} \wedge \mathbf{v2}) \cdot (\mathbf{v3} \wedge \mathbf{v4}) = \mathbf{v1} \cdot (\mathbf{v2} \wedge (\mathbf{v3} \wedge \mathbf{v4}))$?
Let
$\mathbf{v1} = q1*(c11*\mathbf{u1} + c12*\mathbf{u2} + c13*\mathbf{u3})$
$\mathbf{v2} = q2*(c21*\mathbf{u1} + c22*\mathbf{u2} + c23*\mathbf{u3})$
$\mathbf{v3} = q3*(c31*\mathbf{u1} + c32*\mathbf{u2} + c33*\mathbf{u3})$
$\mathbf{v4} = q4*(c41*\mathbf{u1} + c42*\mathbf{u2} + c43*\mathbf{u3})$
Then
$\mathbf{v1} \wedge \mathbf{v2} = q1*q2*((c12*c23 - c13*c22)*\mathbf{u1}$
$\qquad\qquad\qquad + (c13*c21 - c11*c23)*\mathbf{u2}$
$\qquad\qquad\qquad + (c11*c22 - c12*c21)*\mathbf{u3})$
$\mathbf{v3} \wedge \mathbf{v4} = q3*q4*((c32*c43 - c33*c42)*\mathbf{u1}$
$\qquad\qquad\qquad + (c33*c41 - c31*c43)*\mathbf{u2}$
$\qquad\qquad\qquad + (c31*c42 - c32*c41)*\mathbf{u3})$
so
$(\mathbf{v1} \wedge \mathbf{v2}) \cdot (\mathbf{v3} \wedge \mathbf{v4}) = q1*q2*q3*q4$
$\qquad\qquad *((c12*c23 - c13*c22)*\mathbf{u1}$
$\qquad\qquad + (c13*c21 - c11*c23)*\mathbf{u2}$
$\qquad\qquad + (c11*c22 - c12*c21)*\mathbf{u3})$
$\qquad\qquad \cdot ((c32*c43 - c33*c42)*\mathbf{u1}$
$\qquad\qquad + (c33*c41 - c31*c43)*\mathbf{u2}$
$\qquad\qquad + (c31*c42 - c32*c41)*\mathbf{u3})$

$$= (c12*c23 - c13*c22)*(c32*c43 - c33*c42)$$
$$+ (c13*c21 - c11*c23)*(c33*c41 - c31*c43)$$
$$+ (c11*c22 - c12*c21)*(c31*c42 - c32*c41)$$

while
(v1•v3)*(v2•v4) − (v1•v4)*(v2•v3)
$$= q1*q2*q3*q4$$
$$*((c11*c31 + c12*c32 + c13*c33)$$
$$*(c21*c41 + c22*c42 + c23*c43)$$
$$- (c11*c41 + c12*c42 + c13*c43)$$
$$*(c21*c31 + c22*c32 + c23*c33)).$$

Einstein: "They don't look the same."

Newton: "Wait. Let's expand then into single addends.
For **(v1∧v2)•(v3∧v4)**
$$(c12*c23 - c13*c22)*(c32*c43 - c33*c42)$$
$$+ (c13*c21 - c11*c23)*(c33*c41 - c31*c43)$$
$$+ (c11*c22 - c12*c21)*(c31*c42 - c32*c41)$$
$$= (c12*c23*c32*c43$$
$$+ c13*c22*c33*c42$$
$$-c12*c23*c33*c42$$
$$- c13*c22*c32*c43$$
$$+ c13*c21*c33*c41$$
$$+ c11*c23*c31*c43$$
$$-c13*c21*c31*c43$$
$$- c11*c23*c33*c41$$
$$+ c11*c22*c31*c42$$
$$+ c12*c21*c32*c41$$
$$-c11*c22*c32*c41$$
$$- c12*c21*c31*c42$$

while for **(v1•v3)*(v2•v4) − (v1•v4)*(v2•v3)**
$$((c11*c31 + c12*c32 + c13*c33)$$
$$*(c21*c41 + c22*c42 + c23*c43)$$
$$- (c11*c41 + c12*c42 + c13*c43)$$
$$*(c21*c31 + c22*c32 + c23*c33))$$
$$= c11*c31*c21*c41$$
$$+ c11*c31*c22*c42$$
$$+ c11*c31*c23*c43$$
$$+ c12*c32*c21*c41$$
$$+ c12*c32* c22*c42$$
$$+ c12*c32*c23*c43$$
$$+ c13*c33*c21*c41$$
$$+ c13*c33*c22*c42$$
$$+ c13*c33*c23*c43$$
$$- c11*c41*c21*c31$$
$$- c11*c41*c22*c32$$
$$- c11*c41*c23*c33$$
$$- c12*c42*c21*c31$$

$$- c12*c42*c22*c32$$
$$- c12*c42*c23*c33$$
$$- c13*c43)*c21*c31$$
$$- c13*c43)* c22*c32$$
$$- c13*c43)*c23*c33$$

Einstein: "You have 12 summands in the first compilation and 18 in this one."

Newton: "But six of them cancel; so this final compilation also reduces to 12 summands which are
(v1•v3)*(v2•v4) − (v1•v4)*(v2•v3)
$$= c11*c31*c22*c42$$
$$+ c11*c31*c23*c43$$
$$+ c12*c32*c21*c41$$
$$+ c12*c32*c23*c43$$
$$+ c13*c33*c21*c41$$
$$+ c13*c33*c22*c42$$
$$- c11*c41*c22*c32$$
$$- c11*c41*c23*c33$$
$$- c12*c42*c21*c31$$
$$- c12*c42*c23*c33$$
$$- c13*c43*c21*c31$$
$$- c13*c43* c22*c32$$
Now check these."

Einstein: "They all match! I'm amazed. We've traded geometry for bookkeeping, and the bookkeeping is easier."

Breton: "So Newton now you have a couple more entries for your table."

Newton: "Only one."

Breton: "Actually two. Remember we proved earlier
v1•(v2∧(v3∧v4)) = (v1•v3)*(v2•v4) − (v1•v4)*(v2•v3)
so we also know
$$(v1 \wedge v2) \cdot (v3 \wedge v4) = v1 \cdot (v2 \wedge (v3 \wedge v4)).$$

With that Newton added the following entries to his table.

Defined: four at a time	
$(v1 \wedge v2) \cdot (v3 \wedge v4)$ $= ((v1 \cdot v3)*(v2 \cdot v4)$ $ - (v1 \cdot v2)*(v3 \cdot v4)$	
$(v1 \wedge v2) \cdot (v3 \wedge v4)$ $= v1 \cdot (v2 \wedge (v3 \wedge v4))$	
$(v1 \wedge v2) \cdot (v3 \wedge v4)$ $+ (v1 \wedge v3) \cdot (v4 \wedge v2)$ $+ (v1 \wedge v4) \cdot (v2 \wedge v3) = 0$	

Einstein: "You've added still more entries."

Newton: "And I could have easily added still others. For instance since
$$(v1 \wedge v2) \cdot (v3 \wedge v4) = ((v1 \cdot v3)*(v2 \cdot v4) - (v1 \cdot v2)*(v3 \cdot v4)$$
then
$$(v1 \wedge v3) \cdot (v4 \wedge v2) = ((v1 \cdot v4)*(v3 \cdot v2) - (v1 \cdot v3)*(v4 \cdot v2)$$
and
$$(v1 \wedge v4) \cdot (v2 \wedge v3) = ((v1 \cdot v2)*(v4 \cdot v3) - (v1 \cdot v4)*(v2 \cdot v3)$$
each of which is proved by a mere substitution of labels.
Sum them together; you will find each positive summand matched by a negative summand."

Einstein: "Other possibilities exist. How about $(v1 \wedge v2) \wedge (v3 \wedge v4)$?"

Breton: "That's already solved. Remember
$$v \wedge (v2 \wedge v3) = (v \cdot v3)*v2 - (v \cdot v2)*v3$$
from above? By simply relabeling
$$v2 \text{ as } v3$$
$$v3 \text{ as } v4$$
$$v \text{ as } v1 \wedge v2$$
the equation becomes
$$(v1 \wedge v2) \wedge (v3 \wedge v4) = ((v1 \wedge v2) \cdot v4)*v3 - ((v1 \wedge v2) \cdot v3)*v2$$
which we can rewrite as
$$(v1 \wedge v2) \wedge (v3 \wedge v4) = v4 \cdot (v1 \wedge v2)*v3 - v3 \cdot (v1 \wedge v2)*v2$$
and also since
$$(v1 \wedge v2) \wedge v = (v1 \cdot v)*v2 - (v2 \cdot v)*v1$$
$$(v1 \wedge v2) \wedge (v3 \wedge v4) = v1 \cdot (v3 \wedge v4)*v2 - v2 \cdot (v3 \wedge v4)*v1.$$
So using neither geometry or bookkeeping you can add these identities to your table Newton."

Einstein, with enthusiasm: "How about $v1 \wedge (v2 \wedge (v3 \wedge v4))$ or $((v1 \wedge v2) \wedge v3) \wedge v4$?"

Newton: "Let me try these. We know
$$v_1 \wedge (v_2 \wedge v) = (v_1 \bullet v)*v_2 - (v_1 \bullet v_2)*v$$
so
$v_1 \wedge (v_2 \wedge (v_3 \wedge v_4))$
$$= (v_1 \bullet (v_3 \wedge v_4))*v_2 - (v_1 \bullet v_2)*(v_3 \wedge v_4)$$
and likewise since we know
$(v \wedge v_3) \wedge v_4 = (v \bullet v_4)*v_3 - (v_3 \bullet v_4)*v((v_1 \wedge v_2) \wedge v_3) \wedge v_4$
$$= ((v_1 \wedge v_2) \bullet v_4))*v_3 - (v_3 \bullet v_4)*(v_1 \wedge v_2).$$
Easy. Indeed, I begin to see relationships between our results. Look
$v_1 \wedge (v_2 \wedge (v_3 \wedge v_4))$
$$= (v_1 \bullet (v_3 \wedge v_4))*v_2 - (v_1 \bullet v_2)*(v_3 \wedge v_4)$$
so
$v_1 \wedge ((v_3 \wedge v_4) \wedge v_2)$
$$= (v_1 \bullet v_2)*(v_3 \wedge v_4) - (v_1 \bullet (v_3 \wedge v_4))*v_2$$
which can be rewritten
$v_1 \wedge ((v_2 \wedge v_3) \wedge v_4)$
$$= (v_1 \bullet v_3)*(v_4 \wedge v_2) - (v_1 \bullet (v_4 \wedge v_2))*v_3$$
$$= (v_1 \bullet (v_2 \wedge v_4))*v_3 - (v_1 \bullet v_3)*(v_2 \wedge v_4)$$
and likewise
$((v_1 \wedge v_2) \wedge v_3) \wedge v_4$
$$= ((v_1 \wedge v_2) \bullet v_4))*v_3 - (v_3 \bullet v_4)*(v_1 \wedge v_2)$$
so
$(v_3 \wedge (v_1 \wedge v_2)) \wedge v_4$
$$= (v_3 \bullet v_4)*(v_1 \wedge v_2) - ((v_1 \wedge v_2) \bullet v_4))*v_3$$
which can be rewritten
$(v_1 \wedge (v_2 \wedge v_3)) \wedge v_4$
$$= (v_1 \bullet v_4)*(v_2 \wedge v_3) - ((v_2 \wedge v_3) \bullet v_4))*v_1.$$

Einstein: "Two different results for the same thing."

Breton: "Two roads leading to the same destination. We have come again to the distinction between 'is' and 'equals'. The two different expressions are not the same, but they are equal as vectors. Remember how earlier we noted that {2+2} and {3+1} are different numerical expressions with the same value. Now we see two different vectorial expressions having the same vectorial value."

Newton: "Another breathtaking intellectual vista."

Breton: "Just as with numbers, the many expressions for same value lead to equations and eventually to an algebra. We can note here that from the above we have
$(v_1 \bullet (v_2 \wedge v_4))*v_3 - (v_1 \bullet v_3)*(v_2 \wedge v_4)$
$$= (v_1 \bullet v_4)*(v_2 \wedge v_3) - ((v_2 \wedge v_3) \bullet v_4))*v_1$$
which becomes on transposing,
$(v_1 \bullet (v_2 \wedge v_4))*v_3 + ((v_2 \wedge v_3) \bullet v_4))*v_1$
$$= (v_1 \bullet v_4)*(v_2 \wedge v_3) + (v_1 \bullet v_3)*(v_2 \wedge v_4).$$

Einstein: "Let me observe that with four vectors we can insert three different multiplications."

Breton: "But not all of these are legitimate. For instance, $v1 \cdot v2 \cdot v3 \cdot v4$ makes no sense."

Einstein: "While $v1 \cdot (v2 \cdot v3 * v4)$ does. So let us continue with other possibilities."

Newton: "Some are easy enough. Let me write some for you.
$$v1 \cdot (v2 \cdot v3 * v4) = (v1 \cdot v4) * (v2 \cdot v3)$$
$$v1 \cdot (v2 * v3) \cdot v4 = (v1 \cdot v2) * (v3 \cdot v4)$$
$$= v1 \cdot (v2 * v4) \cdot v3$$
$$= v2 \cdot (v1 * v3) \cdot v4$$
$$= v2 \cdot (v1 * v4) \cdot v3.$$

Einstein: "You've done well with the combination {inner, inner, outer}. Now try {inner, outer, outer}!"

Newton: "All =right
$$(v1 \cdot v2) * (v3 * v4) = (v1 \cdot (v2 * v3)) * v4$$
$$= v3 * (v1 \cdot v2) * v4$$
$$= (v3 * v1) \cdot (v2 * v4)$$
$$= v3 * (v2 \cdot v1) * v4$$
$$= (v3 * v2) \cdot (v1 * v4)$$
$$(v1 * v2) \cdot (v3 * v4) = (v2 \cdot v3) * (v1 * v4).$$

Einstein: "The last equation is not obvious."

Newton: "It's simple enough. For any vector v
$$v \cdot (v1 * v2) \cdot (v3 * v4) = (v \cdot v1) * v2 \cdot (v3 * v4)$$
$$= (v \cdot v1) * (v2 \cdot v3) * v4$$
$$= (v2 \cdot v3) * (v \cdot v1) * v4$$
$$= (v2 \cdot v3) * v \cdot (v1 * v4)$$
so the action of any vector on $(v1 * v2) \cdot (v3 * v4)$ is the same as that on $(v2 \cdot v3) * (v1 * v4)$."

Breton: "So it appears that the inner and outer products act together, inner products producing scalars and outer products producing transformations."

Einstein, clearly enjoying his leadership: "How about the combination {inner, cross, outer}?"

Newton: "Not too difficult
$$v1 \cdot ((v2 \wedge v3) * v4) = (v1 \cdot (v2 \wedge v3)) * v4$$
$$= (v2 \cdot (v3 \wedge v1)) * v4$$
$$= (v3 \cdot (v1 \wedge v2)) * v4$$
$$= ((v1 \wedge v2) \cdot v3) * v4$$
$$= ((v2 \wedge v3) \cdot v1) * v4$$
$$= ((v3 \wedge v1) \cdot v2) * v4$$
$$= v2 \cdot ((v3 \wedge v1) * v4)$$

$$= v3 \cdot ((v1 \wedge v2) * v4)$$
which are all variations of the triple scalar product as well as.
$$(v1 \wedge v2) \cdot (v3 * v4) = (v1 \cdot (v2 \wedge v3)) * v4.$$
In contrast
$$v1 \wedge (v2 \cdot (v3 * v4)) = (v2 \cdot v3) * (v1 \wedge v4).$$
From the vector triple product with a transposition we have
$$(v1 \cdot v2) * (v3 \wedge v4) = (v1 \wedge v3) \cdot (v4 * v2) + (v1 \wedge v3) \wedge (v4 \wedge v2).$$
Reversing the order we obtain again from the scalar triple product
$$\begin{aligned}(v1 * (v2 \wedge v3)) \cdot v4 &= v1 * ((v2 \wedge v3) \cdot v4) \\ &= v1 * ((v3 \wedge v4) \cdot v2) \\ &= v1 * ((v4 \wedge v2) \cdot v3) \\ &= v1 * (v2 \cdot (v3 \wedge v4)) \\ &= v1 * (v3 \cdot (v4 \wedge v2)) \\ &= v1 * (v4 \cdot (v2 \wedge v3)).\end{aligned}$$

Breton: "The effort to prove triple products is paying dividends."

Einstein pushing on with determination: "The combination {vector, cross, outer} is missing."

Breton: "For a good reason. The operation
$$v \wedge (v1 * v2)$$
is not defined. It is worthwhile noting that
$$v1 \cdot (v2 * v3) - (v2 * v3) \cdot v1 = (v2 \wedge v3) \wedge v1,$$
not 0 generally, and that $(v1 \wedge v2) \cdot (v3 * v4)$ does not equal $v1 \wedge (v2 \cdot (v3 * v4))$.
Now Newton, would you be good enough to put all of these into a table to which we can easily refer."

Newton: "Gladly."

Axiomatic	Comments
$v1 + v2 = v3$	closure
$v1 + v2$ $= (q1 * c11 + q2 * c21) * u1$ $ + (q1 * c12 + q2 * c22) * u2$ $ + (q1 * c13 + q2 * c23) * u3$	Reference origin
$v1 + (v2 + v3) = (v1 + v2) + v3$	association
$q * v1 = v2$	Scalar multiplication
Two at a time	
$v1 \cdot v2 = v2 \cdot v1$	Inner product
$v1 \cdot v2 = q1 * q2$ $ * (c11 * c21$ $ + c12 * c22$ $ + c13 * c23)$	Reference origin
$b * v1 \cdot c * v2 = b * c * (v1 \cdot v2)$	

$\text{abs}(\mathbf{v1} \cdot \mathbf{v2}) \le \text{abs}(\mathbf{v1}) * \text{abs}(\mathbf{v2})$	
$\mathbf{v1} \wedge \mathbf{v2} = -(\mathbf{v2} \wedge \mathbf{v1})$ $= ((-\mathbf{v2}) \wedge \mathbf{v1})$ $= (\mathbf{v2} \wedge (-\mathbf{v1}))$	cross product
$\mathbf{u1} \wedge \mathbf{u2} \equiv \mathbf{u3}$ $\mathbf{u2} \wedge \mathbf{u3} \equiv \mathbf{u1}$ $\mathbf{u3} \wedge \mathbf{u1} \equiv \mathbf{u2}$	Reference origin
$\mathbf{v1} \wedge \mathbf{v2}$ $= q1*q2$ $*((c12*c23 - c13*c22)*\mathbf{u1}$ $+ (c13*c21 - c11*c23)*\mathbf{u2}$ $+ (c11*c22 - c12*c21)*\mathbf{u3})$	Reference origin
$\mathbf{v1} \wedge \mathbf{v1} = 0$	
$\mathbf{v1} \cdot (\mathbf{v1} \wedge \mathbf{v2}) = \mathbf{v2} \cdot (\mathbf{v1} \wedge \mathbf{v2}) = 0$	
$(b*\mathbf{v1}) \wedge (c*\mathbf{v2}) = b*c*(\mathbf{v1} \wedge \mathbf{v2})$	
$\text{abs}(\mathbf{v1} \wedge \mathbf{v2}) \le \text{abs}(\mathbf{v1})*\text{abs}(\mathbf{v2})$	
$(b*\mathbf{v1})*(c*\mathbf{v2}) = b*c*(\mathbf{v1}*\mathbf{v2})$	
$\mathbf{v1}*\mathbf{v2} = q1*q2*\mathbf{uv1}*\mathbf{uv2}$	Reference origin
$\mathbf{v1}*\mathbf{v2} = q1*q2$ $*(c11*c12*\mathbf{u1}*\mathbf{u1}$ $+ c11*c22*\mathbf{u1}*\mathbf{u2}$ $+ c11*c23*\mathbf{u1}*\mathbf{u3}$ $+ c12*c12*\mathbf{u2}*\mathbf{u1}$ $+ c12*c22*\mathbf{u2}*\mathbf{u2}$ $+ c12*c23*\mathbf{u2}*\mathbf{u3}$ $+ c13*c12*\mathbf{u3}*\mathbf{u1}$ $+ c13*c22*\mathbf{u3}*\mathbf{u2}$ $+ c13*c23*\mathbf{u3}*\mathbf{u3})$	Reference origin
$(\text{abs}(\mathbf{v1})*\text{abs}(\mathbf{v2}))^2$ $= (\text{abs}(\mathbf{v1} \wedge \mathbf{v2}))^2$ $+ (\text{abs}(\mathbf{v1} \cdot \mathbf{v2}))^2$	
Three at a time	
$\mathbf{v1} \cdot (\mathbf{v2}+\mathbf{v3}) = \mathbf{v1} \cdot \mathbf{v2} + \mathbf{v1} \cdot \mathbf{v3}$	
$\mathbf{v1} \wedge (\mathbf{v2}+\mathbf{v3}) = \mathbf{v1} \wedge \mathbf{v2} + \mathbf{v1} \wedge \mathbf{v3}$	
$\mathbf{v1}*(\mathbf{v2}+\mathbf{v3}) = \mathbf{v1}*\mathbf{v2} + \mathbf{v1}*\mathbf{v3}$	
$\mathbf{v1} \cdot (\mathbf{v2} \wedge \mathbf{v3}) = \mathbf{v2} \cdot (\mathbf{v3} \wedge \mathbf{v1})$ $= \mathbf{v3} \cdot (\mathbf{v1} \wedge \mathbf{v2})$ $= (\mathbf{v1} \wedge \mathbf{v2}) \cdot \mathbf{v3}$ $= (\mathbf{v2} \wedge \mathbf{v3}) \cdot \mathbf{v1}$ $= (\mathbf{v3} \wedge \mathbf{v1}) \cdot \mathbf{v2}$	Scalar triple product
$\mathbf{v1} \cdot (\mathbf{v2}*\mathbf{v3}) = (\mathbf{v1} \cdot \mathbf{v2})*\mathbf{v3}$	transformation

v1∧(v2∧v3) = (v1•v3)*v2 − (v1•v2)*v3 = v3•(v1*v2) − v3*(v1•v2) = v1•(v3*v2 − v2*v3)	Vector triple product
v1∧(v2∧v3) − (v1∧v2)∧v3 = v2•(v3*v1 − v1*v3)	

v1∧(v2∧v3)
  + v2∧(v3∧v1)
  + v3∧(v1∧v2) = 0	

Four at a time	

(v1∧v2)•(v3∧v4)
   = v1•(v2∧(v3∧v4))
   = ((v1•v3)*(v2•v4)
       − (v1•v2)*(v3•v4)
	{cross,inner,cross}

(v1∧v2)•(v3∧v4)
  + (v1∧v3)•(v4∧v2)
  + (v1∧v4)•(v2∧v3) = 0
	{cross,inner,cross}

(v1∧v2)∧(v3∧v4)
   = v4•(v1∧v2)*v3
     −v3•(v1∧v2)*v2
	{cross,cross,cross}

((v1∧v2)∧v3)∧v4
   = ((v1∧v2)•v4))*v3
      − (v3•v4)*(v1∧v2)
	{cross,cross,cross}

(v1∧(v2∧v3))∧v4
   = (v1•v4)*(v2∧v3)
      − ((v2∧v3)•v4))*v1
{cross,cross,cross}
	{cross,cross,cross}

v1∧(v2∧(v3∧v4))
   = (v1•(v3∧v4))*v2
      − (v1•v2)*(v3∧v4)
	{cross,cross,cross}

v1∧((v2∧v3)∧v4)
   = (v1•v4)*(v2∧v3)
     −(v1•(v2∧v3))*v4
	{cross,cross,cross}

(v1•(v2∧v4))*v3
   + ((v2∧v3)•v4))*v1
   = (v1•v4)*(v2∧v3)
      + (v1•v3)*(v2∧v4)
	{inner,cross,outer}

v1•(v2•v3*v4) = (v1•v4)*(v2•v3)	{inner,inner,outer}

v1•(v2*v3)•v4
   = (v1•v2)*(v3•v4)
   = v1•(v2*v4)•v3
   = v2•(v1*v3)•v4
	{inner,outer,inner}

$= v2 \cdot (v1 * v4) \cdot v3$	
$(v1 \cdot v2) * (v3 * v4)$ $= (v1 \cdot (v2 * v3)) * v4$ $= v3 * (v1 \cdot v2) * v4$ $= (v3 * v1) \cdot (v2 * v4)$ $= v3 * (v2 \cdot v1) * v4$ $= (v3 * v2) \cdot (v1 * v4)$	{inner,outer,outer}
$(v1 * v2) \cdot (v3 * v4)$ $= (v2 \cdot v3) * (v1 * v4)$	{outer,inner,outer}
$v1 \cdot ((v2 \wedge v3) * v4)$ $= (v1 \cdot (v2 \wedge v3)) * v4$ $= (v2 \cdot (v3 \wedge v1)) * v4$ $= (v3 \cdot (v1 \wedge v2)) * v4$ $= ((v1 \wedge v2) \cdot v3) * v4$ $= ((v2 \wedge v3) \cdot v1) * v4$ $= ((v3 \wedge v1) \cdot v2) * v4$ $= v2 \cdot ((v3 \wedge v1) * v4)$ $= v3 \cdot ((v1 \wedge v2) * v4)$	{inner,cross,outer}
$(v1 \wedge v2) \cdot (v3 * v4)$ $= (v1 \cdot (v2 \wedge v3)) * v4$	{cross,inner,outer}
$v1 \wedge (v2 \cdot (v3 * v4))$ $= (v2 \cdot v3) * (v1 \wedge v4)$	{cross,inner,outer}
$(v1 \cdot v2) * (v3 \wedge v4)$ $= (v1 \wedge v3) \cdot (v4 * v2)$ $+ (v1 \wedge v3) \wedge (v4 \wedge v2)$	{inner,outer,cross}
$(v1 * (v2 \wedge v3)) \cdot v4$ $= v1 * ((v2 \wedge v3) \cdot v4)$ $= v1 * ((v3 \wedge v4) \cdot v2)$ $= v1 * ((v4 \wedge v2) \cdot v3)$ $= v1 * (v2 \cdot (v3 \wedge v4))$ $= v1 * (v3 \cdot (v4 \wedge v2))$ $= v1 * (v4 \cdot (v2 \wedge v3))$	{outer,cross,inner}

Breton: "You have constructed a remarkable table, Newton, a veritable armory of intellectual tools. We have climbed much higher up our mountain."

Newton: "How fascinating that each of these vectorial expressions corresponds to a geometrical theorem whose proof might be very difficult indeed."

Einstein: "This wonderful facility arises from the way we defined the origin. The vectorial origin corresponds to the zero of the quotient numbers. It must be special indeed."

Newton: "Just as my illustrious ancestor said and furthermore he both named this origin and claimed it as an absolute location."

Breton: "Whatever the claim made about a physical origin, I am ready to prove that the origin of the axiomatic set of vectors is completely arbitrary."

Newton, taken aback: "I shall oppose your reasoning with every fiber of my being. Success for you would devastate not only my illustrious ancestor, but also most of classical physics."

Einstein: "Just as my illustrious ancestor claimed."

Breton: "Illustrious ancestors aside, shall we just proceed reasonably and logically without appeals to previous authority?"

Einstein, enthusiastically: "Please proceed with your proof."

Breton: "Let me first describe the logic of the proof. Suppose a given origin has been chosen. If any other vector in the axiomatic set of vectors can replace the given origin, then the choice of origin is arbitrary."

Newton: "What do you mean by replace?"

Breton: "That all the elementary functions of vectors referred to the initial origin can be re-expressed in terms of the second origin."

Newton: "But the expressions will different."

Breton: "Very likely."

Newton: "Different expressions mean the origins are not arbitrary. One may be preferable to another."

Breton: "Still both expressions are valid. There is no intrinsic reason for choosing one over the other."

Einstein, adding: "Similar to look–alike functions where two different expressions have the same value. So we will have, if Breton can prove his contention, two different descriptions of the same thing, which is not at all the same as two different descriptions of two different things. Let's get on with the proof."

Breton: "First let me label the given origin **v0**. Secondly, let me take another vector **v1** as a candidate for the new origin. Then for any vector **v**

$$\mathbf{v} - \mathbf{v1} = (\mathbf{v} - \mathbf{v0}) - (\mathbf{v1} - \mathbf{v0})$$
$$\mathbf{v} - \mathbf{v0} = (\mathbf{v} - \mathbf{v1}) - (\mathbf{v0} - \mathbf{v1})$$

Now label any vector referenced to **v0** as origin as
$$v|v0 \equiv v - v0$$
and the same vector referenced to **v1** as origin as
$$v|v1 \equiv v - v1$$
Then the equations can be written as
$$v|v1 = v|v0 - v1|v0$$
$$v|v0 = v|v1 - v0|v1.$$
So any vector can be expressed as well with either **v0** or **v1** taken as the origin."

Newton: "Not so fast. You have shown either vector can serve as the zero of the axiomatic set of vectors, but not necessarily as origin. The origin incorporates an orientation, remember?"

Breton: "How could I forget? Let
$$A = u1 \bullet ur1 + u2 \bullet ur2 + u3 \bullet ur3$$
where **u1**, **u2**, **u3** are the orientation of **v0** and **ur1**, **ur2**, **ur3** are three mutually orthogonal directions which will serve as the orientation of **v1**.
   Then **A** has an inverse
$$A^{-1} = ur1 \bullet u1 + ur2 \bullet u2 + ur3 \bullet u3$$
since $A \bullet A^{-1} = I$, the identity transformation.
Now for any vector **v**
$$(v - v0) \bullet I = ((v - v1) - (v0 - v1)) \bullet A \bullet A^{-1}$$
$$= (v - v1) \bullet A \bullet A^{-1} - (v0 - v1) \bullet A \bullet A^{-1}$$
Now let
$$v|v1 = (v - v1) \bullet A$$
to indicate both a change in position and reorientation.
Then
$$v|v1 = (v|v0 - v1|v0) \bullet A$$
for **v1** as origin, and
$$v|v0 = (v|v1 - v0|v1) \bullet A^{-1}$$
for **v0** as origin.
Thus **v1** is as suitable to serve as reference as **v0**, both for locations and orientations, and thus as origin for any vector **v**."

Einstein: "Exactly. No location is absolute; any location is described as only relative to the origin. But suppose the origin itself is moving."

Breton: "How would you know it is moving?"

Newton, recovering: "If the vector **v** and the origin were both moving identically, then it would appear that **v** is not moving."

Breton: "So **v** can appear as moving or not moving."

Einstein: "Breton, you've led us into another thicket. How will you get us out of this one."

Breton: "Consider the origin as defining a given perspective. So long as the origin remains constant, the perspective remains the same. But if the origin changes, the perspective changes. For instance, were the origin rotated 180 degrees, what was first perceived as forward, would then be perceived as backward. Similarly, if a vector is perceived as moving referred to the first origin, the movement would appear different referred to a different origin. If the second origin were moving, the perspective might change enough to make the vector appear to have stopped moving."

Einstein, conclusively: "So there is no true location or true motion?"

Breton: "No absolutely true location or motion exists physically, only location and motion referred to an origin."

Einstein: "That is: all location and motion are relative."

Breton: "It is not their existence which is relative, but only our perception of them."

Newton, interjecting heatedly: "I disagree vehemently. When my illustrious ancestor sought to understand the forces operating in the solar system, he used the perspective of the motion relative to the sun as origin. On this basis he formulate the axioms of his *Principia*. If these axioms are absolutely true, and let me add they result in his definition of gravity, then the sun must be an absolute location."

Breton: "And if the sun is not an absolute location, then his axioms are not absolutely true."

Newton: "Then classical mechanics is not absolutely true."

Einstein: "Of course, just as my illustrious ancestor said."

Breton: "Well now, we have departed no little way from considering vector sets and their proposed algebra. Let us take one step at a time, lest we fall on our faces. I have merely established that position and movement of vectors are described relative to an origin. For a different origin, the description may change. As Einstein pointed out, these descriptions appear as look-alike functions."

Both Einstein and Newton say almost simultaneously: "What next?"

Breton: "We have constructed intellectually an axiomatic set of vectors which can be added and multiplied, but we shall not have an algebra until we learn how to divide vectors."

Einstein: "Ask any mathematician, vectors cannot be divided."

Breton, continuing stoically: "We found that multiplication in the axiomatic set of vectors needed to be defined. So let us try to define division in the axiomatic set of vectors without reference to opinion of others which may be erroneous."

Newton: "How?"

Breton: "Remember in the set of quotient numbers, Q, for every quotient number, q, except 0, we could find another quotient number q1 such that $q * q1 = 1$. Thus we could define $1/q \equiv q1$ and so we were able to define division in Q using reciprocal quotient numbers.

Now any vector $\mathbf{v} = q(\mathbf{v})*\mathbf{uv}$. We already have a reciprocal for $q(\mathbf{v})$ so we only need deal with **uv**."

Newton, suddenly seeing the analogy: "And we already know $\mathbf{uv} \bullet \mathbf{uv} = 1$. So for any vector $\mathbf{v}$
$$\mathbf{v} \bullet ((1/q(\mathbf{v}))*\mathbf{uv})) = q(\mathbf{v})*\mathbf{uv} \bullet ((1/q(\mathbf{v}))*\mathbf{uv})) = 1.$$

Breton: "By your leave let us label the vector $((1/q(\mathbf{v}))*\mathbf{uv}))$ as $\mathbf{qd}(\mathbf{v})$. So you see dear Einstein, division in the axiomatic set of vectors is not only possible, but even easy. For every vector $\mathbf{v}$ there exists a reciprocal vector $\mathbf{qd}(\mathbf{v})$ such that
$$\mathbf{v} \bullet \mathbf{qd}(\mathbf{v}) = 1.$$

Einstein: "Except for **0**."

Breton: "Similar to the quotient numbers."

Newton: "This is wonderful. We are scaling up the mountain. The axiom of scalar multiplication contained implicitly the possibility of scalar division."

Einstein: "But for the axiomatic set of vectors other multiplications are defined in addition to the inner product."

Breton: "You are astute Einstein. So there must be other kinds of division for the axiomatic set of vectors. Let me then propose a formal definition of two different reciprocal vectors."

Definition (reciprocal vectors)
Given
$$\mathbf{v} = q(\mathbf{v})*\mathbf{uv} = q(\mathbf{v})*(c1*\mathbf{u1}+c2*\mathbf{u2}+c3*\mathbf{u3})$$
then
$$\mathbf{qd}(\mathbf{v}) \equiv \mathbf{uv}/q(\mathbf{v}) = (c1*\mathbf{u1}+c2*\mathbf{u2}+c3*\mathbf{u3})/q(\mathbf{v})$$
is called the **directional reciprocal vector** of **v**,
and
$$\mathbf{q}(\mathbf{v}) \equiv \mathbf{q}(\mathbf{uv})/q(\mathbf{v}) = (\mathbf{u1}/c1+\mathbf{u2}/c2+\mathbf{u3}/c3)/q(\mathbf{v})$$
is called the **general reciprocal vector** of **v**,
end of definition

Notice that
$$\mathbf{v} \bullet \mathbf{qd}(\mathbf{v}) = 1$$
while
$$\mathbf{v} \bullet \mathbf{q}(\mathbf{v}) = 3.$$

Newton: "Are there others?"

Solutions of vector algebraic equations

Breton: "Yes, but this will do for now. Now that we have assembled an algebra for our axiomatic set of vectors, let us solve a few algebraic equations. What is the solution for a vector **x** where
$$\mathbf{x} \cdot \mathbf{v1} = q1$$
and **v1** is a given vector and q1 is a given quotient number?"

Solutions of equations involving inner products

Einstein, boldly: "Let's break out the equation a little more. Say
$$\mathbf{x} = q(\mathbf{x}) * \mathbf{ux}$$
and
$$\mathbf{v1} = q(\mathbf{v1}) * \mathbf{uv1}$$
Then
$$q(\mathbf{x}) * \mathbf{ux} \cdot \mathbf{uv1} = q1/q(\mathbf{v1})$$
For *any* direction **ux**
$$q(\mathbf{x}) = q1/(q(\mathbf{v1}) * \mathbf{ux} \cdot \mathbf{uv1})$$
so that many $q(\mathbf{x})$ are possible.

For *any* magnitude $q(\mathbf{x})$
$$\mathbf{ux} \cdot \mathbf{uv1} = q1/q(\mathbf{v1}) * q(\mathbf{x}))$$
so that **ux** may be taken from any circle on the unit sphere whose cosine with **uv1** equals $q1/q(\mathbf{v1}) * q(\mathbf{x}))$.

So there is no solution to the equation."

Breton: "An admirable development to a false conclusion. Let me give you a solution,
$$\mathbf{x1} = q1 * \mathbf{qd}(\mathbf{v1})$$
since $q1 * \mathbf{qd}(\mathbf{v1}) \cdot \mathbf{v1} = q1 * 1 = q1$."

Newton: "So Einstein's argument shows not there are no solutions, but there are a great many with differing directions and magnitudes."

Breton: "The truth of the matter is sprouting. We can go further. Let **x1** and **x2** be two different solutions. Then
$$(\mathbf{x1} - \mathbf{x2}) \cdot \mathbf{v1} = q1 - q1 = 0$$
so the difference between any two solution vectors is orthogonal to **v1**. Consequently, the entire set of solutions describes a plane orthogonal to **v1**.

In particular for any solution vector let
$$\mathbf{x} - q1 * \mathbf{qd}(\mathbf{v1}) = q * \mathbf{un}$$
for some quotient number q and **un** a direction orthogonal to **v1**. Then the entire set of solutions may be written as
$$\{\mathbf{x} | \mathbf{x} = q1 * \mathbf{qd}(\mathbf{v1}) + q * \mathbf{un}\}.$$

Einstein: "What's so special about the reciprocal vector $\mathbf{qd}(\mathbf{v1})$ for a solution?"

Breton: "As you have shown, the magnitude of any solution can be written as
$$q(\mathbf{x}) = q1/(q(\mathbf{v1}) * \mathbf{ux} \cdot \mathbf{uv1})$$
so the minimum magnitude is the one which maximizes $\mathbf{ux} \cdot \mathbf{uv1}$."

Newton: "Which is **ux** = **uv1**!"

Breton: "Exactly. Then the unique solution with minimum magnitude is
$$\mathbf{xm} \equiv q(\mathbf{xm}) * \mathbf{uxm} = q1 * \mathbf{uv1}/q(\mathbf{v1})$$
$$= q1 * \mathbf{qd(v1)}.$$

Einstein, regrouping: "Remarkable. Just as with quotient numbers, a divisor can be any of an infinite number of quotient numbers. For instance,
$$3/2 = 3/(4/2) = 3/(6/3)$$
and so on."

Breton, happily acquiescing: "The unity we perceive is indeed remarkable, and intellectually beautiful.

But now let us turn to climb a little higher up our mountain. We have seen outer products as a kind of transformation which changes one vector into another. Because of its utility let me now consider transformations more generally as matrices."

Einstein: "A new word. Define what you mean."

Breton: "Of course. Here is a definition.

Definition (matrix)
  Given
   three vectors, **v1**, **v2**, **v3**.
   an origin designated by **u1**, **u2**, **u3**
  then a matrix **A** is defined as
   $\mathbf{A} \equiv \mathbf{u1} * \mathbf{v1} + \mathbf{u2} * \mathbf{v2} + \mathbf{u3} * \mathbf{v3}$
                                                end of definition
---

The matrix can be represented as an ordered array of vectors, as follows. For
$$\mathbf{v1} = v11*\mathbf{u1} + .v12*\mathbf{u2} + .v13*\mathbf{u3}$$
$$\mathbf{v2} = v21*\mathbf{u1} + .v22*\mathbf{u2} + .v23*\mathbf{u3}$$
$$\mathbf{v3} = v31*\mathbf{u1} + .v32*\mathbf{u2} + .v33*\mathbf{u3}$$
Then **A** may be written as
$$\mathbf{A} = v11*\mathbf{u1}*\mathbf{u1} + v12*\mathbf{u1}*\mathbf{u2} + v13*\mathbf{u1}*\mathbf{u3}$$
$$+ v21*\mathbf{u2}*\mathbf{u1} + v22*\mathbf{u2}*\mathbf{u2} + v23*\mathbf{u2}*\mathbf{u3}$$
$$+ v31*\mathbf{u3}*\mathbf{u1} + v23*\mathbf{u3}*\mathbf{u2} + v33*\mathbf{u3}*\mathbf{u3}$$
that is as the sum of the product of nine scalar coefficients with corresponding unit outer products. The unit outer products can be taken as place-holders in an array and so the matrix can be conveniently symbolized as
$$\mathbf{A} = \begin{bmatrix} v11 & v12 & v13 \\ v21 & v22 & v23 \\ v31 & v32 & v33 \end{bmatrix}$$
You can see that with this symbol the matrix is composed of nine elements arranged in three rows and three columns and implies a given orientation."

Einstein: "How does the matrix fit with our definitions?"

Breton: "Let's start with a simple outer product. Let the direction vectors of the origin's orientation be represented as
$$\mathbf{u1} = (1,0,0)$$
$$\mathbf{u2} = (0,1,0)$$
$$\mathbf{u3} = (0,0,1)$$
and
$$\mathbf{v1} = (v11, v12, v13)$$
Now
$$\mathbf{u1} * \mathbf{v1} = \mathbf{u1} * (v11*\mathbf{u1} + v12*\mathbf{u2} + v13*\mathbf{u3})$$
$$= v11*\mathbf{u1}*\mathbf{u1} + v12*\mathbf{u1}*\mathbf{u2} + v13*\mathbf{u1}*\mathbf{u3}.$$
This same result is expressed in matrix notation as
$$\begin{bmatrix} 1 \\ 0 \\ 0 \end{bmatrix} \cdot \begin{bmatrix} v11 & v12 & v13 \end{bmatrix}$$

where the first element of the column vector is multiplied by each member of the horizontal vector to form the topmost row of the matrix, the second element of the column vector, 0, is multiplied by each member of the horizontal vector to form the middle row of the matrix, and the third element of the column vector, 0, is multiplied by each member of the horizontal vector to form the bottom row of the matrix.

The result becomes
$$\begin{bmatrix} 1 \\ 0 \\ 0 \end{bmatrix} \cdot \begin{bmatrix} v11 & v12 & v13 \end{bmatrix} = \begin{bmatrix} v11 & v12 & v13 \\ 0 & 0 & 0 \\ 0 & 0 & 0 \end{bmatrix}$$

You can see that **u1∗u1** corresponds to position of the first row and first column; that **u1∗u2** corresponds to position of the first row and second column; that **u1∗u3** corresponds to position of the first row and the third column."

Einstein: "What good is all this bookkeeping for?"

Breton: "It does seem complicated, but it matches the complicated process of vector multiplication well. For instance the inner product of two vectors **v1•v2** can be expressed as
$$\begin{bmatrix} v11 & v12 & v13 \end{bmatrix} \cdot \begin{bmatrix} v21 \\ v22 \\ v23 \end{bmatrix} = v11*v21 + v12*v22 + v13*v23$$

So this matrix multiplication comprehends both inner and outer vector multiplications."

Newton: "You have added to the above description by defining matrix multiplication to resemble inner products."

Breton: "True enough. We could say succinctly that in matrix multiplication rows are multiplied with columns as inner products. And if we agree on this definition, then two matrices can be multiplied together, as

$$\begin{bmatrix} v11 & v12 & v13 \\ v21 & v22 & v23 \\ v31 & v32 & v33 \end{bmatrix} \cdot \begin{bmatrix} x11 & x12 & x13 \\ x21 & x22 & x23 \\ x31 & x32 & x33 \end{bmatrix}$$

$= (v11*x11 + v12*x21 + v13*x31)\bullet u1\bullet u1$
$\quad + (v11*x12 + v12*x22 + v13*x32)\bullet u1\bullet u2$
$\quad + (v11*x13 + v12*x23 + v13*x33)\bullet u1\bullet u3$
$\quad + (v21*x11 + v22*x21 + v23*x31)\bullet u2\bullet u1$
$\quad + (v21*x12 + v22*x22 + v23*x32)\bullet u2\bullet u2$
$\quad + (v21*x13 + v22*x23 + v23*x33)\bullet u2\bullet u3$
$\quad + (v31*x11 + v32*x21 + v33*x31)\bullet u3\bullet u1$
$\quad + (v31*x12 + v32*x22 + v33*x32)\bullet u3\bullet u2$
$\quad + (v31*x13 + v32*x23 + v33*x33)\bullet u3\bullet u3.$

Einstein: "If this defines matrix multiplication, then it is ambiguous! A vector in this notation can be either vertical or horizontal. How can that be?"

Before Breton could answer, Einstein continued.

Einstein: "Let's step back a little. We aimed at trying to define an algebra for the axiomatic set of vectors. Now it seems we have not only accomplished that, but also defined an algebra for matrices. Breton, write the operations for matrix algebra explicitly!"

Breton: "Better still let me construct a table showing the algebra of both vectors and matrices. The symbols of the table are defined as

v1 ≡ v1∗**uv1**
v2 ≡ v2∗**uv2**

With the origin taken as reference these vectors are further specified as

v1 ≡ v11∗**u1** + v12∗**u2** + v13∗**u3**)
v1 ≡ v1∗(c11∗**u1** + c12∗**u2** + c13∗**u3**)
v2 ≡ v21∗**u1** + v22∗**u2** + v23∗**u3**)
v2 ≡ v2∗(c21∗**u1** + c22∗**u2** + c23∗**u3**)

Orientation reference for vectors	
Addition	
v1+v2	(v11+v21)∗**u1** (v12+v22)∗**u2** (v13+v23)∗**u3**
Subtraction	
v1−v2	(v11−v21)∗**u1** (v12−v22)∗**u2** (v13−v23)∗**u3**

Multiplication	
$v1 \cdot v2$	$(v11*v21)$ $+ (v12*v22)$ $+ (v13*v23)$
$v1 \wedge v2$	$(v12*v23-v13*v22)*u1$ $+(v13*v21-v11*v23)*u2$ $+(v11*v22-v12*v21)*u3$
$v1*v2$	$(v11*v21)*u1*u1$ $+ (v11*v22)*u1*u2$ $+ (v11*v23)*u1*u3$ $+ (v12*v21)*u2*u1$ $+ (v12*v22)*u2*u2$ $+ (v12*v23)*u2*u3$ $+ (v13*v21)*u3*u1$ $+ (v13*v22)*u3*u2$ $+(v13*v23)*u3*u3$
Division	
$v1 \cdot qd(v2)$	$(v1/v2)$ $*(c11*c21$ $+c12*c22$ $+c13*c23)$
$v1 \wedge q(v2)$	$(v1/v2)$ $*((c12/c23-c13/c22)*u1$ $+(c13/c21-c11/c23)*u2$ $+(c11/c22-c12/c21)*u3)$
$v1*qd(v2)$	$(v1/v2)$ $*(c11*c21)*u1*u1$ $+ (c11*c22)*u1*u2$ $+ (c11*c23)*u1*u3$ $+ (c12*c21)*u2*u1$ $+ (c12*c22)*u2*u2$ $+ (c12*c23)*u2*u3$ $+ (c13*c21)*u3*u1$ $+ (c13*c22)*u3*u2$ $+(c13*c23)*u3*u3$

Einstein: "You have added many things there, Breton. For instance you have defined a cross multiplication."

Breton: "And also an outer product multiplication. The cross multiplication is just a restatement of our earlier definition when we first discussed the origin. The outer product multiplication comes from our earlier discussion of matrix multiplication.

With the origin taken as reference, the symbols for the matrix table are defined for two matrices **A1** and **A2** specified as

$$A1 \equiv u1*v1 + u2*v2 + u3*v3$$
$$A2 \equiv u1*x1 + u2*x2 + u3*x3$$
$$v1 \equiv v1*(c11*u1 + c12*u2 + c13*u3)$$
$$v2 \equiv v2*(c21*u1 + c22*u2 + c23*u3)$$
$$v3 \equiv v3*(c31*u1 + c32*u2 + c33*u3)$$

Here is the table for matrix operations.

Origin reference for matrices	
Addition	
A1+A2	u1*(v1+x1) + u2*(v2+x2) + u3*(v3+x3)
Subtraction	
A1−A2	u1*(v1−x1) + u2*(v2−x2) + u3*(v3−x3)
Multiplication	

A1•A2	(u1•v1*x1•u1
	+u2•v1*x2•u1
	+u3•v1*x3•u1)
	             *u1*u1
	+(u1•v1*x1•u2
	+u2•v1*x2•u2
	+u3•v1*x3•u2
	             *u1*u2
	+(u1•v1*x1•u3
	+u2•v1*x2•u3
	+u3•v1*x3•u3
	             *u1*u3
	+(u1•v2*x1•u1
	+u2•v2*x2•u1
	+u3•v2*x3•u1)
	             *u2*u1
	+(u1•v2*x1•u2
	+u2•v2*x2•u2
	+u3•v2*x3•u2
	             *u2*u2
	+(u1•v2*x1•u3
	+u2•v2*x2•u3
	+u3•v2*x3•u3
	             *u2*u3
	+(u1•v3*x1•u1
	+u2•v3*x2•u1
	+u3•v3*x3•u1)
	             *u3*u1

		+(u1•v3*x1•u2 +u2•v3*x2•u2 +u3•v3*x3•u2 　　　*u3*u2 +(u1•v3*x1•u3 +u2•v3*x2•u3 +u3•v3*x3•u3 　　　*u3*u3
	Division	
	A1•A2⁻¹	(u1•v1*(x2∧x3)•u1 +u2•v1*(x3∧x1)•u1 +u3•v1*(x1∧x2)•u1) 　　　*u1*u1/det[A2] +(u1•v1*(x2∧x3)•u2 +u2•v1*(x3∧x1)•u2 +u3•v1*(x1∧x2)•u2) 　　　*u1*u2/det[A2] +(u1•v1*(x2∧x3)•u3 +u2•v1*(x3∧x1)•u3 +u3•v1*(x1∧x2)•u3) 　　　*u1*u3/det[A2] +(u1•v2*(x2∧x3)•u1 +u2•v2*(x3∧x1)•u1 +u3•v2*(x1∧x2)•u1) 　　　*u2*u1/det[A2] +(u1•v2*(x2∧x3)•u2 +u2•v2*(x3∧x1)•u2 +u3•v2*(x1∧x2)•u2) 　　　*u2*u2/det[A2] +(u1•v2*(x2∧x3)•u3 +u2•v2*(x3∧x1)•u3 +u3•v2*(x1∧x2)•u3) 　　　*u2*u3/det[A2] +(u1•v3*(x2∧x3)•u1 +u2•v3*(x3∧x1)•u1 +u3•v3*(x1∧x2)•u1) 　　　*u3*u1/det[A2] +(u1•v3*(x2∧x3)•u2 +u2•v3*(x3∧x1)•u2 +u3•v3*(x1∧x2)•u2) 　　　*u3*u2/det[A2] +(u1•v3*(x2∧x3)•u3 +u2•v3*(x3∧x1)•u3 +u3•v3*(x1∧x2)•u3) 　　　*u3*u3/det[A2]

Einstein: "You use the same symbols for vector and matrix operations but they mean different operations completely."

Breton: "The ambiguity is resolved by context. The matrices are always capitalized, while the vectors are symbolized with lower case letters."

Einstein: "What is this undefined symbol det[**A**]?

Breton: "The symbol *det* refers to a matrix function. Let me list several useful ones and their definitions. For a matrix defined as

$$\mathbf{A} = \begin{bmatrix} a11 & a12 & a13 \\ a21 & a22 & a23 \\ a31 & a32 & a33 \end{bmatrix}$$

$$\equiv \mathbf{u1*a1 + u2*a2 + u3*a3}.$$

tr[**A**] ≡ **u1•A•u1 + u2•A•u2 + u3•A•u3**
g[**A**] ≡ a11*u1 + a22*u2 + a33*u3
c[**A**] ≡ (a23-a32)*u1 + (a31-a13)*u2 + (a12-a21)*u3
T[**A**] ≡ a1*u1+a2*u2+a3*u3

$$= \begin{bmatrix} a11 & a21 & a33 \\ a12 & a22 & a32 \\ a13 & a23 & a33 \end{bmatrix}$$

det[**A**] ≡ **a1•(a2∧a3) = a2•(a3∧a1) = a3•(a1∧a2)**
  = a11*a22*a33 + a12*a23*a31 + a13*a21*a32
    −a11*a23*a32 − a12*a21*a33 − a13*a22*a31

A^{-1} ≡ ((a2∧a3)*u1 + (a3∧a1)*u2 + (a1∧a2)*u3)/det[**A**]

where the *ai* are the row vectors of **A**; tr[**A**], called the **trace** of **A**, is the sum of the diagonal elements of **A**; g[**A**], called the **diagonal matrix operator**, transforms the matrix into a vector specified by its diagonal elements; c[**A**], called the **curl matrix operator**, transforms the matrix into a vector specified by its off-diagonal elements; T[**A**], called the **transpose** of **A**, interchanges the rows and columns of **A**; det[**A**], called the **determinant** of **A**, is the "volume" of the constituent vectors of **A**; **A**$^{-1}$ is called the **inverse** of **A** since

$$\mathbf{A} \cdot [\mathbf{A}^{-1}] = [\mathbf{A}^{-1}] \cdot \mathbf{A} = \mathbf{I}.$$

Only matrices with non-zero determinants have inverses. For such matrices it is easy to see that

$$T[\mathbf{A}^{-1}] = [T[\mathbf{A}]]^{-1}.$$

For any matrix **A**

det[T[**A**]] = det[**A**]
c[**A**] = −c[T[**A**]].

Newton, somewhat impatiently: "We have algebras. Let's solve more equations."

Solution of equations involving cross products

Einstein: "So can we also solve vector equations involving other multiplications than the inner product? For instance, if
$$\mathbf{x} \wedge \mathbf{v1} = \mathbf{v2}$$
with **v1** and **v2** known, what is **x**?"

Breton: "First may I try something a little easier. How about
$$(\mathbf{v1} \wedge \mathbf{v2}) \cdot \mathbf{x} = q$$
where **v1** and **v2** are given vectors and q a given scalar."

Newton: "That's not so difficult. The equation is a scalar triple product. The cross product **v1**∧**v2** equals
$$qv1*qv2*\sin(\text{angle}(\mathbf{v1},\mathbf{v2})*\mathbf{un}(\mathbf{v1},\mathbf{v2})$$
where **un** is a unit vector orthogonal to both **v1** and **v2**. The problem can then be rewritten as
$$qv1*qv2*\sin(\text{angle}(\mathbf{v2},\mathbf{v2})*\mathbf{un}(\mathbf{v2},\mathbf{v2}) \cdot \mathbf{x} = q$$
the set of whose solutions are already known from our earlier discussion of **v1** • **x** = q. The minimum solution is
$$\mathbf{x} = q*\mathbf{qd}(\mathbf{un}(\mathbf{v1},\mathbf{v2}))/(qv1*qv2*\sin(\text{angle}(\mathbf{v1},\mathbf{v2}))$$
where **qd** is the directional quotient vector."

Breton: "Nicely done. Not so difficult but neither a solution which might be easily guessed at."

Einstein: "Now let's try the solutions to
$$(\mathbf{x} \wedge \mathbf{v1}) \cdot \mathbf{v2} = q.$$

Newton: "Much more difficult. Breton, have you any suggestions."

Breton: "I'd like to introduce another matrix definition which could open a different path to the solution. Remember Einstein suggested that cross multiplication might be expressed with a matrix. Let me now define that matrix."

Definition (curl vector operator)
  Given
$$\mathbf{v} = v*(c1*\mathbf{u1} + c2*\mathbf{u2} + c3*\mathbf{u3})$$
  then

$$C(\mathbf{v}) \equiv v*(-c3*\mathbf{u1}*\mathbf{u2} + c2*\mathbf{u1}*\mathbf{u3}$$
$$+c3*\mathbf{u2}*\mathbf{u1} - c1*\mathbf{u2}*\mathbf{u3}$$
$$-c2*\mathbf{u3}*\mathbf{u1} + c1*\mathbf{u3}*\mathbf{u2})$$

is called the **curl vector operator**.
                                        end of definition

Einstein: "Why do you call it an operator?"

Breton: "The curl vector operator can be written as a matrix
$$C(\mathbf{v}) = \mathbf{v} * \begin{bmatrix} 0 & -c3 & c2 \\ c3 & 0 & -c1 \\ -c2 & c1 & 0 \end{bmatrix}$$
The determinant of $C(\mathbf{v})$ is zer0 so it has no inverse; neither can it be represented as an outer product. We have here an example of different kind of transformation."

Einstein: "How does this help to solve our problem?"

Breton: "First let me define a related matrix function."

Definition (curl matrix function)
Given
$$\mathbf{v1} = v1*(c11*\mathbf{u1}+c12*\mathbf{u2}+c13*\mathbf{u3})$$
$$\mathbf{v2} = v2*(c21*\mathbf{u1}+c22*\mathbf{u2}+c23*\mathbf{u3})$$

then

$$c(v1*v2) \equiv \mathbf{v1} \wedge \mathbf{v2}$$

is called the **curl matrix function**.
end of definition

The curl vector operator and the curl matrix function are related as follows:
$$\mathbf{v1} \wedge \mathbf{v2} = c(v1*v2) = \mathbf{v1} \bullet C(\mathbf{v2})$$
They will likely find use when dealing with cross products."

Einstein: "So let us see if they help with our problem
$$(\mathbf{x} \wedge \mathbf{v1}) \bullet \mathbf{v2} = q!$$

Breton: "We can rewrite the problem as
$$\mathbf{x} \bullet C(\mathbf{v1}) \bullet \mathbf{v2} = q$$
which can be expanded into
$$v1*v2*\mathbf{x} \bullet \begin{bmatrix} 0 & -c13 & c12 \\ c13 & 0 & -c11 \\ -c12 & c11 & 0 \end{bmatrix} \bullet \begin{bmatrix} c21 \\ c22 \\ c23 \end{bmatrix} = q$$
Now let's perform the matrix multiplication on the right to obtain
$$v1*v2*\mathbf{x} \bullet ((-c22*c13 + c23*c12)*\mathbf{u1}$$
$$+(c21*c13 - c23*c11)*\mathbf{u2}$$
$$+(-c21*c12 + c22*c11)*\mathbf{u3})$$
$$= q$$
which is just our familiar solution of $\mathbf{v} \bullet \mathbf{x} = q$ in a different garb."

Newton: "So the solution of $(\mathbf{x} \wedge \mathbf{v1}) \cdot \mathbf{v2} = q$ is
$$\mathbf{x} = q*\mathbf{qd}((-c22*c13 + c23*c12)*\mathbf{u1}$$
$$+ (c21*c13 - c23*c11)*\mathbf{u2}$$
$$+ (-c21*c12 + c22*c11)*\mathbf{u3})/(v1*v2)$$
Not an easy solution to guess at!"

Breton: "Just one of the solutions. An infinite number more exist which are related to the minimum one you have written."

Einstein, looking to refocus the friends: "Now can we address the problem I first proposed? What is the solution to $\mathbf{x} \wedge \mathbf{v1} = \mathbf{v2}$?"

Newton: "Let me try. The problem can be restated as
$$\mathbf{x} \cdot \mathbf{C(v1)} = \mathbf{v2}$$
so $\mathbf{x}$ can be found by simply inverting $\mathbf{C(v1)}$!"

Breton, objecting: "Except that its determinant equals zero and so $\mathbf{C(v1)}$ has no inverse."

Newton, resignedly: "Then the equation has no solutions!"

Breton, plodding forward: "If $\mathbf{v2} = \mathbf{0}$, then $\mathbf{x} = \mathbf{0}$ is a solution."

Newton: "True enough. So are there other solutions?"

Breton: "Let me try.
$$\mathbf{x} \wedge \mathbf{v1} = \mathbf{v2}$$
$$x*\mathbf{ux} \wedge \mathbf{uv1} = v2*\mathbf{uv2}/v1$$
Some solutions are readily apparent. If $\mathbf{v2} = \mathbf{0}$, but not $\mathbf{v1}$, then $\mathbf{x} = \mathbf{0}$ is the only solution. If $\mathbf{v1} = \mathbf{0}$, but not $\mathbf{v2}$, then no solution for $\mathbf{x}$ exists. If both $\mathbf{v2} = \mathbf{0}$ and $\mathbf{v1} = \mathbf{0}$, then $\mathbf{x}$ may be any vector at all."

Einstein: "Where does that leave us?"

Breton: "With the knowledge that solutions not only depend on the directions of $\mathbf{v1}$ and $\mathbf{v2}$, but also their magnitudes."

Einstein, concluding: "Sometimes it's easy, any vector will do; other times it's impossible, no vector will do."

Breton, resolutely plodding forward: "Try thinking about it this way. Suppose $\mathbf{x}$ is a solution. Then $\mathbf{v2}$ must be orthogonal to both $\mathbf{v1}$ and $\mathbf{x}$. So solutions only exist for some $\mathbf{v2}$, but not all. The restriction on $\mathbf{v2}$, then is
$$\mathbf{v2} \cdot \mathbf{v1} = 0$$
From our now familiar solutions to $\mathbf{x} \cdot \mathbf{v} = q$ we know that the only solutions for $\mathbf{v2}$ lie in a plane orthogonal to $\mathbf{v1}$. For any other $\mathbf{v2}$, no solution exists."

Einstein: "Brilliant. Then what is the solution for a given restricted $\mathbf{v2}$?"

Breton: "If **v2•v1** = 0, then let us choose **v2** = qv2∗**un(v1)** where **un(v1)** is orthogonal to **v1**. Then we can rewrite our equation as
$$\mathbf{x} \wedge \mathbf{v1} = qv2*\mathbf{un(v1)}$$
Now we see that **x** has to be orthogonal to **un(v1)** as well. So then both
$$\mathbf{x} \wedge \mathbf{v1} = qx* qv1*\sin(\text{angle}(\mathbf{x},\mathbf{v1}))*\mathbf{un(v1)}$$
$$= qv2*\mathbf{un(x,v1)}$$
define a solution."

Newton: "So
$$qx = qv2/(qv1*\sin(\text{angle}(\mathbf{x},\mathbf{v1})))$$
and
$$\mathbf{un(v1)} = \mathbf{un(x,v1)}.$$

Einstein, triumphantly: "Which still does not define qx since it depends on angle(**x,v1**)!"

Breton: "What it does define is the entire set of solutions. Any solution for **x** must satisfy simultaneously
$$\mathbf{x} \wedge \mathbf{v1} = qx*qv1*\sin(\text{angle}(\mathbf{x},\mathbf{v1}))*\mathbf{un(x,v1)}$$
and
$$\mathbf{x} \wedge \mathbf{v1} = qv2*\mathbf{uv2}$$
All solutions lie in a plane orthogonal to **uv2**, but not all such vectors are solutions, but only those who satisfy
$$qx*\sin(\text{angle}(\mathbf{x},\mathbf{v1})) = qv2/qv1$$
A curve in the plane orthogonal to **v2** designates the entire set of solutions. The whole set of solutions is thus
$$\{\mathbf{x} | \mathbf{x} = (qv2/(qv1*\sin(\text{angle}(\mathbf{x},\mathbf{v1}))))*\mathbf{un(v1,v2)}\}.$$
Among these solutions there is one which minimizes qx, namely the one that maximizes sin(angle(**x,v1**)), specifically the one for which sin(angle(**x,v1**)) = 1. For this minimum solution
$$\mathbf{x} = (qv2/qv1)*\mathbf{un(v1,v2)}.$$

Newton: "Magnificent!!"

Breton: "Similar to the directional quotient vector for inner products, we can define a directional quotient vector for cross products as follows:

Definition (directional quotient vector for cross products)
Given
$$\mathbf{v1} = v1*\mathbf{uv1}$$
$$\mathbf{v2} = v2*\mathbf{uv2}$$
$$\mathbf{un(v1,v2)} = \mathbf{uv1} \wedge \mathbf{uv2}$$
then
$$\mathbf{qd(v1,v2)} \equiv v1*\mathbf{un(v1,v2)}/v2$$
end of definition

Newton, seeing the implications: "Then
$$qd(u1,u2) = u3$$
$$qd(u2,u3) = u1$$
$$qd(u3,u1) = u2$$

Breton: "Exactly."

Solutions of equations involving outer products

Einstein: "How about outer products?"

Breton: "The problem can formulated in any of three ways.
$$x \bullet v1 * v2 = v3$$
$$v1 \bullet x * v2 = v3$$
$$v1 \bullet v2 * x = v3.$$

Newton: "The first two formulations are identical since $x \bullet v1 = v1 \bullet x$."

Breton: "I stand corrected. There are only two possible formulations. Let us start with the first one.
$$x \bullet v1 * v2 = q(v3) * v2 = v3$$
so **v3** must have the same direction as **v2**. Furthermore
$$x \bullet v1 * q(v2) = q(v3).$$

Newton: "Whose solutions for **x** we already know, including a minimum one."

Breton: "Which we can write as
$$x = q(v3) \bullet qd(v1)/q(v2).$$

Einstein: "So let us proceed to $v1 \bullet v2 * x = v3$."

Breton: "Now the direction of **x** must be the same as that of **v3**. So we may rewrite the problem as
$$v1 \bullet v2 * q(x) * uvx = v3 * uv3$$
from which we see
$$q(x) = v3/v1 \bullet v2$$
and $\quad uvx = uv3.$

Einstein: "So for outer products there is only one unique answer. For others, many solutions exist. Newton would you create a table showing these differences."

To this request Newton quickly produced the following table.

Equation	given	restrictions	solution	minimum
$x \cdot v1 = q1$	v1,q1	none	plane	$x = q1 * qd(v1)$
$(v1 \wedge v2) \cdot x = q1$	v1,v2,q1	none	plane	$x = q1$ $*qd(un(v1,v2))$ $/(qv1*qv2$ $*sin(angle(v2,v2))$
$(x \wedge v1) \cdot v2 = q1$	v1,v2,q1	$v2 \cdot v1 = 0$	plane	$x = q1$ $*qd((c23*c12$ $-c22*c13)*u1$ $+(c21*c13$ $-c23*c11)*u2$ $+(c22*c11$ $-c21*c12)*u3)$ $/(v1*v2)$
$x \wedge v1 = v2$	v1, v2	$v2 \cdot v1=0$; $x \cdot un(v1)=0$	curve in plane	$x = (qv2/qv1)$ $*un(v1v2)$
$x \cdot v1 * v2 = v3$	v1, v2, v3	$uv2 = uv3$	plane	$x = v3*qd(v1)/v2$
$v1 \cdot v2 * x = v3$	v1, v2, v3	$uvx = uv3$	unique	$x = v3/(v1 \cdot v2)$

Newton: "So now we kow how to solve vector equations."

Solutions to matrix equations

Einstein: "How about matrix equations?"

Breton: "Since we have formed an algebra of matrices, we should be able to solve matrix equations too."

Einstein: "Sounds like a promise. Deliver!"

Breton: "Let's start with some matrix, **A**, a given vector **v**, and a matrix equation
$$x \cdot A = v$$
and ask for the unknown vector **x**."

Newton, ingenuously: "That's easy. Simply find the inverse of the matrix. Then
$$x = v \cdot A^{-1}$$

Breton: "That will do for matrices with inverses. What about those without inverses, those whose determinants equal zero."

Einstein: "We've seen just this case with the outer product. If
$$A = v1 * v2$$
an outer product, it has no inverse, but $x \cdot v1 * v2$ equals any vector parallel to **v2**. So **v** must be a vector parallel to **v2**; otherwise there is no solution."

Newton, stubbornly: "Show me that outer products have no inverse."

Einstein: "If they did, then the solution would be the unique solution you just demonstrated."

Newton, insisting: "Show me that the determinant of any outer product equals zero."

Breton, ever the peacemaker: "All right, let me do it. Please pay close attention to the manipulations.
   Let **v1** =v1∗(c11∗**u1**+c12∗**u2**+c13∗**u3**
   Let **v2** =v2∗(c21∗**u1**+c22∗**u2**+c23∗**u3**
then
v1∗v2 = v1∗v2∗(c11∗c12∗**u1**∗**u1**
            + c11∗c22∗**u1**∗**u2**
            + c11∗c23∗**u1**∗**u3**
            +c21∗c12∗**u2**∗**u1**
            + c21∗c22∗**u2**∗**u2**
            + c21∗c23∗**u2**∗**u3**
            +c31∗c12∗**u3**∗**u1**
            + c31∗c22∗**u3**∗**u2**
            + c31∗c23∗**u3**∗**u3**)
      = v1∗v2∗(**u1**∗(c11∗c12∗**u1**
            + c11∗c22∗**u2**
            + c11∗c23∗**u3**)
         +**u2**∗(c21∗c12∗**u1**
            + c21∗c22∗**u2**∗**u2**
            + c21∗c23∗**u2**∗**u3**)
         +**u3**∗(c31∗c12∗**u1**
            + c31∗c22∗**u2**
            +c31∗c23∗**u3**))
Therefore
det[**v1∗v2**]=(c11∗c12∗**u1**+ c11∗c22∗**u2**+ c11∗c23∗**u3**)
         ∧(c21∗c12∗**u1**+ c21∗c22∗**u2**+ c21∗c23∗**u3**)
         •(c31∗c12∗**u1**+ c31∗c22∗**u2**+ c31∗c23∗**u3**)
      = (c11∗c12∗ c21∗c22∗**u3** − c11∗c12∗ c31∗c23∗**u2**
         −c11∗c22∗ c21∗c12∗**u3** + c11∗c22∗ c21∗c23∗**u1**
         +c11∗c23∗ c31∗c12∗**u2** − c11∗c23∗ c21∗c22∗**u1**)
         •(c31∗c12∗**u1**+ c31∗c22∗**u2**+ c31∗c23∗**u3**).
   = (0∗**u1** +0∗**u2** +0∗**u3**)
         •(c31∗c12∗**u1**+ c31∗c22∗**u2**+ c31∗c23∗**u3**).
   = 0
So any outer product has a determinant equal to zero."

Einstein, appreciatively: "Thank you. Perhaps you can deliver on your promise after all."

Breton: "The promise recognizes distinctions in the set of matrices. A matrix can be seen as a function
               A:V3 → V3
so we can ask functional questions about it. Is an outer product injective or surjective?

Einstein: "I like the terminology *into* or *onto*. Since the outer product maps any vector into a given direction, it must be an *into* function. And also many-to-one."

Breton: "How about a matrix with an inverse?"

Einstein: "That matrix would be both *onto and 1-1.*"

Breton: "Can you prove your assertion?"

Einstein, accepting the challenge: "Let **x1**•**A** = **v1** and **x2**•**A** = **v1** where **x1**≠ **x2**. Then
$$(x1-x2) \cdot A = 0$$
Now if **A** has and inverse
$$(x1-x2) \cdot A \cdot A^{-1} = (x1-x2) = 0 \cdot A^{-1} = 0$$
So **x1** = **x2**, which contradicts my assumption. So **A** as a function must be 1-1.

Next let **v1** be any vector. Then
$$v1 \cdot A^{-1} = x1$$
for some vector **x1**. For **x1** then
$$x1 \cdot A = v1$$
and so **A** as a function must be onto."

Breton: "Well proven. Matrices as functions then may be either 1-1 and onto, or otherwise, that is, not 1-1 not onto. If **A** = [0] for instance, it would map any vector of the domain into the **0** vector of the range. This is an example of a matrix as a constant function."

Newton, looking to reinstate himself in the conversation: "How can we distinguish between the many types of matrices which are not 1-1 and onto?"

Breton: "To examine further the categories of matrices let me offer the following definition.

Definition (null set of a matrix)
  Given
      A, a matrix
  then

      $N(A) \equiv \{v | v$ is a vector such that $v \cdot A = 0\}$

is called the **null subset** of **A**
                    end of definition

Einstein, conclusively: "Then the vector **0** is a member of null subset of any matrix."

Breton: "Correct. If a non-zero vector **v1** is in the null subset, then for any q, q∗**v1** is also in the subset."

Newton: "That would be all the members of a straight line of vectors."

Breton: "And if two non-parallel vectors find themselves in the null subset?"

Newton, quickly answering: "Then any vector in the plane of vectors containing the two vectors would also be in the null subset."

Breton: "What is the null subset of a matrix with an inverse?"

Newton: "Only the vector **0**."

Breton: "So any given matrix can be categorized as one whose null subset is either a vector line, plane, the whole axiomatic set of vectors, or simply the vector **0**. We label these subsets with a function called dimension, whose value are

subset **0**	dimension = 0
a line	dimension = 1
a plane	dimension = 2
V3	dimension = 3

The entire set of matrices are divided into four subsets each characterized by a dimension."

Newton: "Whereas the partitions of quotient numbers had only two characterizations—0 and line."

Breton: "Yes, you see the similarities. Do you remember the definition of restricted subsets? Note that each of these subsets,- lines, planes, or **v3** entire—comes with its restricted algebra."

Einstein: "And the 0 in quotient numbers evolves into the [**0**] of matrices. How do the subsets relate to **V3**?"

Breton: "We already know some answers. For **A**, any matrix with a **0** null subset,
$$A: V3 \to V3$$
propely.
For **A**, any matrix with **V3** as its null subset,
$$N(A) = 0.$$

Newton: "So if you add the dimension of the null subset to the dimension of the image the result is always 3."

Einstein, elaborating: "So for **A**, any matrix with a line of vectors for its null subset, is the image of **V3** a plane?"

Breton: "Could be. Suppose the null space is the line v1•**u1**. Then **A** would map any vector v•(c1•**u1** +c2•**u2** +c3•**u3**) into some other vector v2•**u2** +v3•**u3**. The image of all such vectors would indeed be a plane orthogonal to **u1**."

Einstein: "How about an arbitrary direction?"

Newton, happily interrupting: "Then we might as well chosen the origin to have **u1** as the arbitrary direction with an identical result."

Breton, summarizing: "So again the dimension of the image added to the dimension of the null set equals 3."

Newton, rounding out the topic: "A similar argument shows the image of a matrix with a plane for a null set would have an image of a line of vectors."

Breton: "We give the name **rank** to the dimension of image of **A**. Then we can write for any matrix **A**
$$\text{dimension}(\mathbf{N}(\mathbf{A})) + \text{rank}(\mathbf{A}) = 3.$$

Einstein: "Why not simply say
$$\text{dimension}(\mathbf{N}(\mathbf{A})) + \text{dimension}(\text{image}(\mathbf{A})) = 3?$$

Breton: "Acceptable, of course. It's just a bit more convenient to talk about the solutions of matrix equations for a matrix of rank 2, than one whose image has dimension of 2."

Einstein: "Proceed then to the solutions of matrix equations."

Breton: "Let's start with the equation
$$\mathbf{x} \cdot \mathbf{A} = \mathbf{b}$$
where the matrix **A** and the vector **b** are given. We seek a solution for **x**."

Newton: "We already have solutions for matrices of rank 0 or rank 3. For a matrix of rank 3 only one vector is a solution. For a matrix of rank 0 any vector is a solution, provided **b** = **0**; otherwise there is no solution."

Breton: "So we might suspect solutions for matrices of rank 1 or 2 might have only some but not all vectors for solutions. Let's start with the explicit solution for matrices of rank 3. Let
$$\mathbf{A} = \mathbf{u1}*\mathbf{a1} + \mathbf{u2}*\mathbf{a2} + \mathbf{u3}*\mathbf{a3}$$
where the **ai** are the rows of the matrix. The solution for
$$\mathbf{x} \cdot \mathbf{T[A]} = \mathbf{b}$$
is
$$\mathbf{x} = \mathbf{b} \cdot \mathbf{T[A]}^{-1}$$
Now for **b** = b*(c1***u1**+c2***u2**+c3***u3**)
b·**T[A]**$^{-1}$
  = b*(c1***u1**+c2***u2**+c3***u3**)
     ·(**u1***(**a2**∧**a3**)+**u2***(**a3**∧**a1**)+**u3***(**a1**∧**a2**))/det[**A**]
  = b*((c1*(**a2**∧**a3**)+c2*(**a3**∧**a1**)+c3*(**a1**∧**a2**))/det[**A**].

Einstein, noticing: "You've changed the problem by substituting the transpose for the original matrix. Why?"

Breton: "The original problem calls for multiplying the unknown vector by the columns of the matrix. By substituting the transpose the solution can be stated in terms of the rows of **A**. If the matrix is given, then so too is its transpose."

Einstein, seizing the initiative: "Let's move on to matrices of other ranks."

Breton: "Suppose now a matrix **A** of rank2. Then **N(A)** is a line of vectors."

Einstein, restating the problem commandingly: "For such a matrix, what is the solution for **x** in the equation
$$\mathbf{x} \cdot [\mathbf{A}] = \mathbf{b}$$
where the matrix **A** and the vector **b** are given?"

Breton: "If **b** = **0** then **x** = **0** is a solution."

Newton, interrupting: "What is image of **A**?"

Breton continuing dismissively: "We might note that if **x1** and **x2** are solutions, then **x1−x2** lies in the null set of **A** since
$$(\mathbf{x1-x2}) \cdot \mathbf{A} = \mathbf{x1} \cdot \mathbf{A} - \mathbf{x2} \cdot \mathbf{A} = \mathbf{b} - \mathbf{b} = 0.$$

Newton: "Then for **b** = **0** there must be other solutions than **x** = **0**."

Breton: "You are right! Let **unull** be the direction of the null set. Then
$$\mathbf{unull} \cdot [\mathbf{A}] = 0$$
So **unull** is also a solution for **x** • [**A**] = **0**."

Newton: "Then so is **x** = t•**unull** for any t in Q."

Breton: "It appears then that the solutions for **x** • [**A**] = **0** are any vector in **N(A)**, that is, the set of such solutions is a line of vectors. We might then expect a line of vectors for solutions to **x** • [**A**] = **b** for any **b**."

Newton: "My question about the image of **A** implies asking if there are any restrictions on **b**, or can any vector be in the image of **A**."

Breton, reflecting: "Yours is indeed an interesting question. If **A** is a rank3 matrix, then any **x** in **V** will find an image under **A**.

Newton: "But if **A** has rank equal to 0, then any **x** in **V3** will have only **0** as its image under **A**. So then only **b** = **0** would be allowed. Any other **b** would not have a solution."

Breton, reflecting further: "A matrix of rank2 would take any line {**x** + t• **unull**} into some plane containing **0**; likewise a matrix of rank1 would take any plane {**x** + t1• **unull1** + t2• **unull2**} into some line containing **0**."

Newton: "So then for a rank2 matrix, **b** would have to be in the proper plane and for a rank1 matrix, **b** would have to be in the proper line. Otherwise the equation **x**•[A] = **b** has no solutions."

After a pause Breton continued: "Given two different non-zero vectors **v1** and **v2** the set
$$\{v|v = t1*v1 + t2*v2, t1,t2 \text{ in } Q\}$$
is a plane of vectors. Any vector in the plane can be designated by specifying t1 and t2."

Einstein: "So what is a solution for a matrix of rank2?"

Breton: "Let me start, but you will have to follow closely. Let **v1**, **v2**, and **unull** be three different non-planar vectors with **unull** the direction of **N(A)**. The vectors **v1** and **v2** are called **probe** vectors. Then the solution vector **x** may be represented as
$$x = t1*v1 + t2*v2 + t3*unull$$
for some t1, t2, and t3. So the solution for
$$x \cdot [A] = b$$
devolves into a solution for
$$(t1*v1 + t2*v2 + t3*unull) \cdot A = b.$$

Newton, observing: "So we need only find the scalar quantities, t1, t2, and t3 for a solution."

Breton: "We can also also calculate some components of the equation. For
$$v1 = v11*u1 + v12*u2 + v13*u3$$
$$v2 = v21*u1 + v22*u2 + v23*u3$$
$$v1 \cdot [A] = (v11*a11 + v12*a21 + v13*a31)*u1$$
$$+(v11*a12 + v12*a22 + v13*a32)*u2$$
$$+(v11*a13 + v12*a23 + v13*a33)*u3$$

$$v2 \cdot [A] = (v21*a11 + v22*a21 + v23*a31)*u1$$
$$+(v21*a12 + v22*a22 + v23*a32)*u2$$
$$+(v21*a13 + v22*a23 + v23*a33)*u3.$$

Newton: "That's a porridge of symbols."

Breton: "Which can be easily confused. Why not simplify by using vectorial notation. Let
$$v1 \cdot [A] \cdot u1 = v11*a11 + v12*a21 + v13*a31$$
$$v1 \cdot [A] \cdot u2 = v11*a12 + v12*a22 + v13*a32$$
$$v1 \cdot [A] \cdot u3 = v11*a13 + v12*a23 + v13*a33$$
$$v2 \cdot [A] \cdot u1 = v21*a11 + v22*a21 + v23*a31$$
$$v2 \cdot [A] \cdot u2 = v21*a12 + v22*a22 + v23*a32$$
$$v2 \cdot [A] \cdot u3 = v21*a13 + v22*a23 + v23*a33$$
Then we can write
$$v1 \cdot [A] = v1 \cdot [A] \cdot u1*u1 + v1 \cdot [A] \cdot u2*u2 + v1 \cdot [A] \cdot u3*u3$$
$$v2 \cdot [A] = v2 \cdot [A] \cdot u1*u1 + v2 \cdot [A] \cdot u2*u2 + v2 \cdot [A] \cdot u3*u3$$

so that
$$(t1*v1 + t2*v2 + t3*\text{unull}) \cdot A$$
$$= t1*v1 \cdot A + t2*v2 \cdot A + 0$$
$$= t1*(v1 \cdot [A] \cdot u1*u1 + v1 \cdot [A] \cdot u2*u2 + v1 \cdot [A] \cdot u3*u3)$$
$$+ t2*(v2 \cdot [A] \cdot u1*u1 + v2 \cdot [A] \cdot u2*u2 + v2 \cdot [A] \cdot u3*u3)$$
$$= b1*u1 + b2*u2 + b3*u3.$$

The components for each separate orientation can be isolated as
$$t1*v1 \cdot [A] \cdot u1 + t2*v2 \cdot [A] \cdot u1 = b1$$
$$t1*v1 \cdot [A] \cdot u2 + t2*v2 \cdot [A] \cdot u2 = b2$$
$$t1*v1 \cdot [A] \cdot u3 + t2*v2 \cdot [A] \cdot u3 = b3$$

So we have three different equations which can be solved for two unknowns."

Einstein: "So do it!"

Breton: "We can rewrite the equations as
$$t2 = (b1 - t1*v1 \cdot [A] \cdot u1)/(v2 \cdot [A] \cdot u1)$$
$$t2 = (b2 - t1*v1 \cdot [A] \cdot u2)/(v2 \cdot [A] \cdot u2)$$
$$t2 = (b3 - t1*v1 \cdot [A] \cdot u3)/(v2 \cdot [A] \cdot u3)$$

First let's solve for t1 from the first two equations.
$$(b1 - t1*v1 \cdot [A] \cdot u1)/(v2 \cdot [A] \cdot u1)$$
$$= (b2 - t1*v1 \cdot [A] \cdot u2)/(v2 \cdot [A] \cdot u2)$$

Then if we multiply each side of the equation by $v2 \cdot [A] \cdot u2$
$$(b1 - t1*v1 \cdot [A] \cdot u1)*v2 \cdot [A] \cdot u2/(v2 \cdot [A] \cdot u1)$$
$$= b2 - t1*v1 \cdot [A] \cdot u2$$

Now adding $t1*v1 \cdot [A] \cdot u2$ to each side
$$t1*v1 \cdot [A] \cdot u2 + (b1 - t1*v1 \cdot [A] \cdot u1)*v2 \cdot [A] \cdot u2/(v2 \cdot [A] \cdot u1)$$
$$= b2$$

Next on the left side of the equation, rationalize the second addendum
$$t1*v1 \cdot [A] \cdot u2 + b1*v2 \cdot [A] \cdot u2/(v2 \cdot [A] \cdot u1)$$
$$- t1*v1 \cdot [A] \cdot u1*v2 \cdot [A] \cdot u2/(v2 \cdot [A] \cdot u1)$$
$$= b2$$

Now we have three separate addenda on the left side. So shift the second of these to the right side by subtracting
$$b1*v2 \cdot [A] \cdot u2/(v2 \cdot [A] \cdot u1)$$
to both sides;
$$t1*v1 \cdot [A] \cdot u2 - t1*v1 \cdot [A] \cdot u1*v2 \cdot [A] \cdot u2/(v2 \cdot [A] \cdot u1)$$
$$= b2 - b1*v2 \cdot [A] \cdot u2/(v2 \cdot [A] \cdot u1)$$

Next rationalize the left hand side
$$(t1*v1 \cdot [A] \cdot u2*v2 \cdot [A] \cdot u1$$
$$- t1*v1 \cdot [A] \cdot u1*v2 \cdot [A] \cdot u2)/(v2 \cdot [A] \cdot u1)$$
$$= b2 - b1*v2 \cdot [A] \cdot u2/(v2 \cdot [A] \cdot u1)$$

and factor out the factor t1;
$$(t1*(v1 \cdot [A] \cdot u2*v2 \cdot [A] \cdot u1 - v1 \cdot [A] \cdot u1*v2 \cdot [A] \cdot u2)/(v2 \cdot [A] \cdot u1)$$
$$= b2 - b1*v2 \cdot [A] \cdot u2/(v2 \cdot [A] \cdot u1)$$

Now rationalize the right hand side
$$(t1*(v1 \cdot [A] \cdot u2*v2 \cdot [A] \cdot u1 - v1 \cdot [A] \cdot u1*v2 \cdot [A] \cdot u2)/(v2 \cdot [A] \cdot u1)$$
$$= (b2*v2 \cdot [A] \cdot u1 - b1*v2 \cdot [A] \cdot u2)/(v2 \cdot [A] \cdot u1))$$

and notice the denominators of both sides are the same and so may be eliminated by multiplying both sides by $(v2 \cdot [A] \cdot u1)$.

Then
$$t1*(v1•[A]•u2*v2•[A]•u1 - v1•[A]•u1*v2•[A]•u2)$$
$$= b2*v2•[A]•u1 - b1*v2•[A]•u2$$
Now dividing both sides by
$$(v1•[A]•u2*v2•[A]•u1 - v1•[A]•u1*v2•[A]•u2)$$
we finally reach
$$t1 = (b2*v2•[A]•u1 - b1*v2•[A]•u2)$$
$$/(v1•[A]•u2*v2•[A]•u1 - v1•[A]•u1*v2•[A]•u2).$$

Einstein: "That's a lot of manipulation."

Breton: "That's the genius of algebra. Starting with an equation containing an unknown, we can add, subtract, multiply and divide each side of the equation by equal entities, rationalize either side, substitute equal entities, always keeping the equality between both sides, looking to isolate the unknown. Once isolated, the unknown becomes known from the resulting equation."

Newton: "My illustrious ancestor loved algebraic manipulations."

Einstein: "Algebraic manipulations can be long and arduous."

Breton: "Algebra requires intellectual labor. Lazy persons attempt to finesse the labor through memorization or cultural opinion but they will never attain the certainty and beauty of our results."

With that the trio paused for a reflective moment.

Having precipitated the moment, Breton continued past it.

Breton: "Here's another way of quickly obtaining results. Now that we have found t1, we can easily find t2 as
$$t2 = (b1*v1•[A]•u2 - b2*v1•[A]•u1)$$
$$/(v2•[A]•u1*v1•[A]•u2 - v2•[A]•u2*v1•[A]•u1).$$

Newton: "How can you put down the result for t2 so quickly and easily?"

Breton: "We are using only the first two equations.
$$t1*v1•[A]•u1 + t2*v2•[A]•u1 = b1$$
$$t1*v1•[A]•u2 + t2*v2•[A]•u2 = b2$$
which we can rewrite as
$$t2*v2•[A]•u2 + t1*v1•[A]•u2 = b2$$
$$t2*v2•[A]•u1 + t1*v1•[A]•u1 = b1$$
By substituting t1 for t2 in our result, we see we should also substitute $v2•[A]•u1$ for $v1•[A]•u2$ and vice-versa; $v1•[A]•u1$ for $v2•[A]•u2$ and vice versa; and b2 for b1 and vice versa. This a process called substitution. Try it."

Newton, after of pause of a few minutes, exclaims: "It works!"

Breton: "So our solution becomes.
$$\mathbf{x} = (t1{*}\mathbf{v1} + t2{*}\mathbf{v2} + t3{*}\mathbf{unull})$$
$$= ((b2{*}\mathbf{v2} \cdot [A] \cdot u1 - b1{*}\mathbf{v2} \cdot [A] \cdot u2))$$
$$/(v2 \cdot [A] \cdot u1{*}v1 \cdot [A] \cdot u2 - v2 \cdot [A] \cdot u2{*}v1 \cdot [A] \cdot u1))$$
$${*}\mathbf{v1}$$
$$+ ((b1{*}\mathbf{v1} \cdot [A] \cdot u2 - b2{*}\mathbf{v1} \cdot [A] \cdot u1))$$
$$/(v2 \cdot [A] \cdot u1{*}\, v1 \cdot [A] \cdot u2 - v2 \cdot [A] \cdot u2{*}v1 \cdot [A] \cdot u1))$$
$${*}\mathbf{v2}$$
$$+ t3{*}\mathbf{unull}).$$

Newton: "Not something we could guess at easily."

Einstein, agnostically: "Show directly that **x** is a solution!"

Breton: "All right. Let us check against the original
$$\mathbf{x} \cdot [A] = (t1{*}\mathbf{v1} + t2{*}\mathbf{v2} + t3{*}\mathbf{unull}) \cdot [A].$$
First calculate some of the components. Remember
$$\mathbf{v1} \cdot [A] = v1 \cdot [A] \cdot u1{*}u1 + v1 \cdot [A] \cdot u2{*}u2 + v1 \cdot [A] \cdot u3{*}u3$$
$$\mathbf{v2} \cdot [A] = v2 \cdot [A] \cdot u1{*}u1 + v2 \cdot [A] \cdot u2{*}u2 + v2 \cdot [A] \cdot u3{*}u3$$
so
$$t1{*}\mathbf{v1} \cdot A = (b2{*}\mathbf{v2} \cdot [A] \cdot u1 - b1{*}\mathbf{v2} \cdot [A] \cdot u2)$$
$${*}(v1 \cdot [A] \cdot u1{*}u1 + v1 \cdot [A] \cdot u2{*}u2 + v1 \cdot [A] \cdot u3{*}u3)$$
$$/(v2 \cdot [A] \cdot u1{*}v1 \cdot [A] \cdot u2 - v2 \cdot [A] \cdot u2{*}v1 \cdot [A] \cdot u1)$$
$$t2{*}\mathbf{v2} \cdot A = (b1{*}\mathbf{v1} \cdot [A] \cdot u2 - b2{*}\mathbf{v1} \cdot [A] \cdot u1)$$
$${*}(v2 \cdot [A] \cdot u1{*}u1 + v2 \cdot [A] \cdot u2{*}u2 + v2 \cdot [A] \cdot u3{*}u3)$$
$$/(v2 \cdot [A] \cdot u1{*}v1 \cdot [A] \cdot u1 - v2 \cdot [A] \cdot u2{*}v1 \cdot [A] \cdot u1)$$
and so
$$(t1{*}\mathbf{v1} + t2{*}\mathbf{v2}) \cdot [A]$$
$$= ((b2{*}\mathbf{v2} \cdot [A] \cdot u1 - b1{*}\mathbf{v2} \cdot [A] \cdot u2)$$
$${*}(v1 \cdot [A] \cdot u1{*}u1 + v1 \cdot [A] \cdot u2{*}u2 + v1 \cdot [A] \cdot u3{*}u3)$$
$$+ (b1{*}\mathbf{v1} \cdot [A] \cdot u2 - b2{*}\mathbf{v1} \cdot [A] \cdot u1)$$
$${*}(v2 \cdot [A] \cdot u1{*}u1 + v2 \cdot [A] \cdot u2{*}u2 + v2 \cdot [A] \cdot u3{*}u3))$$
$$/(v2 \cdot [A] \cdot u1{*}v1 \cdot [A] \cdot u2 - v2 \cdot [A] \cdot u2{*}v1 \cdot [A] \cdot u1)$$
$$= (b2{*}\mathbf{v2} \cdot [A] \cdot u1{*}v1 \cdot [A] \cdot u1 - b1{*}\mathbf{v2} \cdot [A] \cdot u2{*}v1 \cdot [A] \cdot u1$$
$$+ b1{*}\mathbf{v1} \cdot [A] \cdot u2{*}v2 \cdot [A] \cdot u1$$
$$- b2{*}\mathbf{v1} \cdot [A] \cdot u1{*}v2 \cdot [A] \cdot u1)$$
$${*}u1$$
$$/(v2 \cdot [A] \cdot u1{*}v1 \cdot [A] \cdot u2 - v2 \cdot [A] \cdot u2{*}v1 \cdot [A] \cdot u1)$$
$$+ (b2{*}\mathbf{v2} \cdot [A] \cdot u1{*}v1 \cdot [A] \cdot u2 - b1{*}\mathbf{v2} \cdot [A] \cdot u2{*}\, v1 \cdot [A] \cdot u2$$
$$+ b1{*}\mathbf{v1} \cdot [A] \cdot u2{*}v2 \cdot [A] \cdot u2$$
$$- b2{*}\mathbf{v1} \cdot [A] \cdot u1{*}v2 \cdot [A] \cdot u2)$$
$${*}u2$$
$$/(v2 \cdot [A] \cdot u1{*}v1 \cdot [A] \cdot u2 - v2 \cdot [A] \cdot u2{*}v1 \cdot [A] \cdot u1)$$
$$+ (b2{*}\mathbf{v2} \cdot [A] \cdot u1{*}v1 \cdot [A] \cdot u3 - b1{*}\mathbf{v2} \cdot [A] \cdot u2{*}\, v1 \cdot [A] \cdot u3$$
$$+ b1{*}\mathbf{v1} \cdot [A] \cdot u3{*}\, v2 \cdot [A] \cdot u3$$
$$- b2{*}\mathbf{v1} \cdot [A] \cdot u1{*}v2 \cdot [A] \cdot u3)$$
$${*}u3$$
$$/(v2 \cdot [A] \cdot u1{*}v1 \cdot [A] \cdot u2 - v2 \cdot [A] \cdot u2{*}v1 \cdot [A] \cdot u1)$$
$$= (-b1{*}\mathbf{v2} \cdot [A] \cdot u2{*}v1 \cdot [A] \cdot u1 + b1{*}\mathbf{v1} \cdot [A] \cdot u2{*}v2 \cdot [A] \cdot u1)$$
$${*}u1$$
$$/(v2 \cdot [A] \cdot u1{*}v1 \cdot [A] \cdot u2 - v2 \cdot [A] \cdot u2{*}v1 \cdot [A] \cdot u1)$$

$$+ (b2*v2\cdot[A]\cdot u1*v1\cdot[A]\cdot u2 - b2*v1\cdot[A]\cdot u1*v2\cdot[A]\cdot u2)$$
$$*u2$$
$$/(v2\cdot[A]\cdot u1*v1\cdot[A]\cdot u2 - v2\cdot[A]\cdot u2*v1\cdot[A]\cdot u1)$$
$$+ (b2*v2\cdot[A]\cdot u1*v1\cdot[A]\cdot u3 - b1*v2\cdot[A]\cdot u2*v1\cdot[A]\cdot u3$$
$$+b1*v1\cdot[A]\cdot u2*v2\cdot[A]\cdot u3$$
$$-b2*v1\cdot[A]\cdot u1*v2\cdot[A]\cdot u3)$$
$$*u3$$
$$/(v2\cdot[A]\cdot u1*v1\cdot[A]\cdot u2 - v2\cdot[A]\cdot u2*v1\cdot[A]\cdot u1)$$
$$= b1*u1$$
$$+ (b2*u2$$
$$+((b2*(v2\cdot[A]\cdot u1*v1\cdot[A]\cdot u3 - v1\cdot[A]\cdot u1*v2\cdot[A]\cdot u3)$$
$$+b1*(v1\cdot[A]\cdot u2*v2\cdot[A]\cdot u3$$
$$-v2\cdot[A]\cdot u2*v1\cdot[A]\cdot u3))$$
$$*u3$$
$$/(v2\cdot[A]\cdot u1*v1\cdot[A]\cdot u2 - v2\cdot[A]\cdot u2*v1\cdot[A]\cdot u1).$$

Einstein: "So your **x** is not a solution!"

Breton: "But it is! The magnitude of the **u3** direction is merely disguised. Do you remember
 $b3 = t1*v1\cdot[A]\cdot u3 + t2*v2\cdot[A]\cdot u3$?
Now that we know t1 and t2 we can write
$b3 = (b2*v2\cdot[A]\cdot u1 - b1*v2\cdot[A]\cdot u2)*v1\cdot[A]\cdot u3$
$/(v1\cdot[A]\cdot u2*v2\cdot[A]\cdot u1 - v1\cdot[A]\cdot u1*v2\cdot[A]\cdot u2)$
$+ (b1*v1\cdot[A]\cdot u2 - b2*v1\cdot[A]\cdot u1)*v2\cdot[A]\cdot u3$
$/(v2\cdot[A]\cdot u1*v1\cdot[A]\cdot u2 - v2\cdot[A]\cdot u2*v1\cdot[A]\cdot u1).$
Compare this with the complicated scalar for **u3**, and see that they are the same. So **x** is indeed a solution!"

Einstein, speechless. Finally replying: "Not so. Suppose **b** = b1***u1**+b2***u2**+0***u3**. Then since your solution is based on the first two equations only, you would come to the same t1 and t2 which would solve for b1 and b2, but for the wrong b3."

Breton: "The vector **b** is given. Designating a different **b** changes the problem. The same probe vectors might not work for a different problem."

Einstein, now curious and turning helpful: "Remember we have only used the first two of three equations which do not explicitly reference b3. What do you think would result if we had used the first and the third equations?"

Newton, somewhat irritated at Einstein's agnosticism,: "We can try substitution. Let
 $t1*v1\cdot[A]\cdot u1 + t2*v2\cdot[A]\cdot u1 = b1$
 $t1*v1\cdot[A]\cdot u3 + t2*v2\cdot[A]\cdot u3 = b3.$
Then this substitution yields
x $= ((b3*v2\cdot[A]\cdot u1 - b1*v2\cdot[A]\cdot u3)$
 $/(v1\cdot[A]\cdot u3* v2\cdot[A]\cdot u1 - v1\cdot[A]\cdot u1*v2\cdot[A]\cdot u3))$
 $*v1$

$$+(b1*v_1 \cdot [A] \cdot u_3 - b_3 * v_1 \cdot [A] \cdot u_1)$$
$$/(v_1 \cdot [A] \cdot u_3 * v_2 \cdot [A] \cdot u_1 - v_1 \cdot [A] \cdot u_1 * v_2 \cdot [A] \cdot u_3))$$
$$*v_2$$
$$+t_3 * unull)$$

which verifies to
$$x = b_1 * u_1$$
$$+((b_3 * (v_2 \cdot [A] \cdot u_1 * v_1 \cdot [A] \cdot u_2 - v_1 \cdot [A] \cdot u_1 * v_2 \cdot [A] \cdot u_2)$$
$$+b_1 * (v_1 \cdot [A] \cdot u_3 * v_2 \cdot [A] \cdot u_2$$
$$-v_2 \cdot [A] \cdot u_3 * v_1 \cdot [A] \cdot u_2))$$
$$*u_2$$
$$/(v_1 \cdot [A] \cdot u_3 * v_2 \cdot [A] \cdot u_1 - v_1 \cdot [A] \cdot u_1 * v_2 \cdot [A] \cdot u_3)$$
$$+ b_3 * u_3.$$

Likewise using the second and third equations,
$$t_1 * v_1 \cdot [A] \cdot u_2 + t_2 * v_2 \cdot [A] \cdot u_2 = b_2$$
$$t_1 * v_1 \cdot [A] \cdot u_3 + t_2 * v_2 \cdot [A] \cdot u_3 = b_3$$

yields
$$x = (b_3 * v_2 \cdot [A] \cdot u_2 - b_2 * v_2 \cdot [A] \cdot u_3)$$
$$/(v_1 \cdot [A] \cdot u_3 * v_2 \cdot [A] \cdot u_2 - v_1 \cdot [A] \cdot u_2 * v_2 \cdot [A] \cdot u_3))$$
$$*v_1$$
$$+(b_2 * v_1 \cdot [A] \cdot u_3 - b_3 * v_1 \cdot [A] \cdot u_2)$$
$$/(v_1 \cdot [A] \cdot u_3 * v_2 \cdot [A] \cdot u_2 - v_1 \cdot [A] \cdot u_2 * v_2 \cdot [A] \cdot u_3))$$
$$*v_2$$
$$+t_3 * unull)$$

which verifies to
$$x = ((b_3 * (v_2 \cdot [A] \cdot u_2 * v_1 \cdot [A] \cdot u_1 - v_2 \cdot [A] \cdot u_1 * v_1 \cdot [A] \cdot u_3)$$
$$+b_2 * (v_1 \cdot [A] \cdot u_3 * v_2 \cdot [A] \cdot u_1 - v_2 \cdot [A] \cdot u_3 * v_1 \cdot [A] \cdot u_1))$$
$$*u_1$$
$$/(v_1 \cdot [A] \cdot u_3 * v_2 \cdot [A] \cdot u_2 - v_1 \cdot [A] \cdot u_2 * v_2 \cdot [A] \cdot u_3))$$
$$+ b_2 * u_2 + b_3 * u_3.$$

Breton: "Different paths to the same destination."

Einstein, incredulous, but reflecting: "So all of these solutions depend on the choice of v_1 and v_2. Choose different probing vectors, and we get different solutions. Let's do an example."

Breton: "All right. Let
$$A = u_1 * u_1 + u_1 * u_2 + u_1 * u_3$$
$$+3 * u_2 * u_1 + 2 * u_2 * u_2 + u_2 * u_3$$
$$+3 * u_3 * u_1 + 3 * u_3 * u_2 + 3 * u_3 * u_3$$
The null subset of A is $\{t_3 * (3 * u_1 - u_3)\}$.
Let
$$b = u_1 + u_2 + u_3.$$
Then $b_1 = 1 = b_2 = b_3$.

Einstein: "How can we tell is b is in the image of A?

Breton: "A good question. If the solution yields a unique vector, then b will be in the image. Let's try with
$$v_1 = u_1$$
$$v_2 = u_2.$$

Then
$$v1 \cdot [A] \cdot u1 = 1$$
$$v1 \cdot [A] \cdot u2 = 1$$
$$v1 \cdot [A] \cdot u3 = 1$$
$$v2 \cdot [A] \cdot u1 = 3$$
$$v2 \cdot [A] \cdot u2 = 2$$
$$v2 \cdot [A] \cdot u3 = 1$$

So from the first and second equations
$$\mathbf{x} = ((b2*v2\cdot[A]\cdot u1 - b1*v2\cdot[A]\cdot u2))$$
$$/(v2\cdot[A]\cdot u1*v1\cdot[A]\cdot u2 - v2\cdot[A]\cdot u2*v1\cdot[A]\cdot u1))$$
$$*u1$$
$$+ ((b1*v1\cdot[A]\cdot u2 - b2*v1\cdot[A]\cdot u1))$$
$$/(v2\cdot[A]\cdot u1* v1\cdot[A]\cdot u2 - v2\cdot[A]\cdot u2*v1\cdot[A]\cdot u1))$$
$$*u2$$
$$+t3*\mathbf{unull}).$$
$$\mathbf{x} = ((1*3-1*2)/(1*3-1*2))*u1$$
$$+((1*1-1*1)/(1*3-1*2))*u2$$
$$+ t3*\mathbf{unull})$$
$$= (u1+ t3*\mathbf{unull})$$

verifying.
$$(u1+ t3*\mathbf{unull}) \cdot [u1*u1+u1*u2+u1*u3]$$
$$= u1 + u2 + u3$$
$$= \mathbf{b}.$$

For the first and third equations,
$$\mathbf{x} = ((b3*v2\cdot[A]\cdot u1-b1*v2\cdot[A]\cdot u3)$$
$$/(v1\cdot[A]\cdot u3* v2\cdot[A]\cdot u1 -v1\cdot[A]\cdot u1*v2\cdot[A]\cdot u3))$$
$$*u1$$
$$+(b1*v1\cdot[A]\cdot u3-b3*v1\cdot[A]\cdot u1)$$
$$/(v1\cdot[A]\cdot u3* v2\cdot[A]\cdot u1 -v1\cdot[A]\cdot u1*v2\cdot[A]\cdot u3))$$
$$*u2$$
$$+t3*\mathbf{unull})$$
$$= ((1*3-1*1)/(1*3-1*1))*u1$$
$$+((1*1-1*1)/(1*3-1*1))*u2$$
$$= (u1+t3*\mathbf{unull})$$

which, as before verifies to **u1 + u2+ u3**, which is **b**.

For the second and third equations,
$$\mathbf{x} = (b3*v2\cdot[A]\cdot u2-b2*v2\cdot[A]\cdot u3)$$
$$/(v1\cdot[A]\cdot u3* v2\cdot[A]\cdot u2 -v1\cdot[A]\cdot u2*v2\cdot[A]\cdot u3))$$
$$*u1$$
$$+(b2*v1\cdot[A]\cdot u3-b3*v1\cdot[A]\cdot u2)$$
$$/(v1\cdot[A]\cdot u3* v2\cdot[A]\cdot u2 -v1\cdot[A]\cdot u2*v2\cdot[A]\cdot u3))$$
$$*u2$$
$$+t3*\mathbf{unull})$$
$$= ((1*2-1*1)/(1*2-1*1))*u1$$
$$+((1*1-1*1)/(1*2-1*1))*u2$$
$$= (u1+t3*\mathbf{unull})$$

which also verifies **u1 + u2 + u3**, which is **b**.
So in this example, all three equations yield the same verifiable solution."

Einstein, grudgingly: "The example is too simple. Let **v1** = **u1** + 2***u2** + **u3**."

Breton: "Then
$$v1 \cdot [A] \cdot u1 = 10$$
$$v1 \cdot [A] \cdot u2 = 8$$
$$v1 \cdot [A] \cdot u3 = 6$$
$$v2 \cdot [A] \cdot u1 = 3$$
$$v2 \cdot [A] \cdot u2 = 2$$
$$v2 \cdot [A] \cdot u3 = 1$$

Now for the first and second equations,
x = ((1*3−1*2)/(8*3 −10*2))*(**u1** +2***u2** + **u3**)
    +((1*8−1*10)/(8*3 −10*2))***u2**
    +t3***unull**)
 = ((**u1**+2***u2**+**u3**)/4) − (**u2**/2) + t3***unull**)
 = ((**u1**+**u3**)/4)+t3***unull**)
Now
((**u1**+**u3**)/4)+t3***unull**) · **A**
 = ((**u1**+**u3**)/4)+t3***unull**) · (**u1**·**u1**+**u1**·**u2**+**u1**·**u3**
                                  +3*u2*u1+2*u2*u2+u2*u3
                                  +3*u3*u1+3*u3*u2+3*u3*u3)
 = (**u1** + **u2**+ **u3**/4 + 3*(**u1** + **u2**+ **u3**/4
which verifies to **u1** + **u2** + **u3**, which is **b**."

Einstein, stubbornly: "Try the first and third equations."

Breton, patiently obliging: "Then
x = ((1*3−1*1)/(6*3 −10*1))*(**u1** +2***u2** + **u3**)
    +((1*6−1*10)/(6*3−10*1))***u2**
 = (2/8)*(**u1**+2***u2** + **u3**)+(−4)/(8)***u2**
x = (**u1**+2***u2**+**u3**)/4−**u2**/2
x = (**u1**+ **u3**)/4
as before."

Einstein, doggedly: "And the second and third equations?"

Breton, patiently, but with a touch of irritation: "Then
x = ((2−1)/(12 −8))***u1**
    +((6−8)/(12 −8))***u2**
x = (1/4)*(**u1**+2***u2** + **u3**)+(−2/4))***u2**
x = (**u1**+ **u3**)/4
as before."

Einstein, yielding: "So the equations yield the same solution." Then, persisting: "Show that the difference between the two solution lies in the null set of **A**."

Breton: "Easily.
$$u1- (u1+ u3)/4 = (4*u1- u1- u3)/4 = (3*u1- u3)/4$$
which is in the null set of **A**."

Einstein, finally yielding: "So we have discovered the solutions to
x· [**A**] = **b** where **A** is a matrix of rank 2."

Then, seizing an initiative to lead, Einstein continued: "If **b** is in **N(A)**, then what is the solution to
$$\mathbf{x} \cdot [A] = \mathbf{b}$$
where the matrix **A** and the vector **b** are given?"

Breton: "Then the solution vector **x** may still be represented as
$$\mathbf{x} = t1*\mathbf{v1} + t2*\mathbf{v2} + t3*\mathbf{unull}$$
for some t1, t2, and t3. So the solution for
$$\mathbf{x} \cdot [A] = \mathbf{b}$$
where now **b** = q1***unull** and so
$$(t1*\mathbf{v1} + t2*\mathbf{v2} + t3*\mathbf{unull}) \cdot A = q*\mathbf{unull}$$
where q is known since
$$\mathbf{b} \cdot \mathbf{unull} = q1*\mathbf{unull} \cdot \mathbf{unull} = q.$$
As before we can also also calculate some components of the equation. For
$$\mathbf{v1} = v11*\mathbf{u1} + v12*\mathbf{u2} + v13*\mathbf{u3}$$
$$\mathbf{v2} = v21*\mathbf{u1} + v22*\mathbf{u2} + v23*\mathbf{u3}$$
$$\mathbf{v1} \cdot [A] = (v11*a11 + v12*a21 + v13*a31)*\mathbf{u1}$$
$$+ (v11*a12 + v12*a22 + v13*a32)*\mathbf{u2}$$
$$+ (v11*a13 + v12*a23 + v13*a33)*\mathbf{u3}$$
$$\mathbf{v2} \cdot [A] = (v21*a11 + v22*a21 + v23*a31)*\mathbf{u1}$$
$$+ (v21*a12 + v22*a22 + v23*a32)*\mathbf{u2}$$
$$+ (v21*a13 + v22*a23 + v23*a33)*\mathbf{u3}.$$
$$\mathbf{v1} \cdot [A] \cdot \mathbf{u1} = v11*a11 + v12*a21 + v13*a31$$
$$\mathbf{v1} \cdot [A] \cdot \mathbf{u2} = v11*a12 + v12*a22 + v13*a32$$
$$\mathbf{v1} \cdot [A] \cdot \mathbf{u3} = v11*a13 + v12*a23 + v13*a33$$
$$\mathbf{v2} \cdot [A] \cdot \mathbf{u1} = v21*a11 + v22*a21 + v23*a31$$
$$\mathbf{v2} \cdot [A] \cdot \mathbf{u2} = v21*a12 + v22*a22 + v23*a32$$
$$\mathbf{v2} \cdot [A] \cdot \mathbf{u3} = v21*a13 + v22*a23 + v23*a33$$
Then we can write
$$\mathbf{v1} \cdot [A] = \mathbf{v1} \cdot [A] \cdot \mathbf{u1}*\mathbf{u1} + \mathbf{v1} \cdot [A] \cdot \mathbf{u2}*\mathbf{u2} + \mathbf{v1} \cdot [A] \cdot \mathbf{u3}*\mathbf{u3}$$
$$\mathbf{v2} \cdot [A] = \mathbf{v2} \cdot [A] \cdot \mathbf{u1}*\mathbf{u1} + \mathbf{v2} \cdot [A] \cdot \mathbf{u2}*\mathbf{u2} + \mathbf{v2} \cdot [A] \cdot \mathbf{u3}*\mathbf{u3}$$
so that
$$(t1*\mathbf{v1} + t2*\mathbf{v2}) \cdot A$$
$$= t1*(\mathbf{v1} \cdot [A] \cdot \mathbf{u1}*\mathbf{u1} + \mathbf{v1} \cdot [A] \cdot \mathbf{u2}*\mathbf{u2} + \mathbf{v1} \cdot [A] \cdot \mathbf{u3}*\mathbf{u3})$$
$$+ t2*(\mathbf{v2} \cdot [A] \cdot \mathbf{u1}*\mathbf{u1} + \mathbf{v2} \cdot [A] \cdot \mathbf{u2}*\mathbf{u2} + \mathbf{v2} \cdot [A] \cdot \mathbf{u3}*\mathbf{u3})$$
$$= q*\mathbf{unull}$$
The components for each separate orientation can be isolated as
$$t1*\mathbf{v1} \cdot [A] \cdot \mathbf{u1} + t2*\mathbf{v2} \cdot [A] \cdot \mathbf{u1} = q*\mathbf{unull} \cdot \mathbf{u1} = b1$$
$$t1*\mathbf{v1} \cdot [A] \cdot \mathbf{u2} + t2*\mathbf{v2} \cdot [A] \cdot \mathbf{u2} = q*\mathbf{unull} \cdot \mathbf{u2} = b2$$
$$t1*\mathbf{v1} \cdot [A] \cdot \mathbf{u3} + t2*\mathbf{v2} \cdot [A] \cdot \mathbf{u3} = q*\mathbf{unull} \cdot \mathbf{u3} = b3.$$

Newton, seeing similarities: "We can write the solutions for t1 and t2 by substitution.
If for
$$t1*\mathbf{v1} \cdot [A] \cdot \mathbf{u1} + t2*\mathbf{v2} \cdot [A] \cdot \mathbf{u1} = b1$$
$$t1*\mathbf{v1} \cdot [A] \cdot \mathbf{u2} + t2*\mathbf{v2} \cdot [A] \cdot \mathbf{u2} = b2$$
$$t1*\mathbf{v1} \cdot [A] \cdot \mathbf{u3} + t2*\mathbf{v2} \cdot [A] \cdot \mathbf{u3} = b3$$
$$t1 = (b2*\mathbf{v2} \cdot [A] \cdot \mathbf{u1} - b1*\mathbf{v2} \cdot [A] \cdot \mathbf{u2})$$
$$/(\mathbf{v1} \cdot [A] \cdot \mathbf{u2}*\mathbf{v2} \cdot [A] \cdot \mathbf{u1} - \mathbf{v1} \cdot [A] \cdot \mathbf{u1}*\mathbf{v2} \cdot [A] \cdot \mathbf{u2}).$$

$$t2 = (b1*\mathbf{v1} \cdot [A] \cdot \mathbf{u2} - b2*\mathbf{v1} \cdot [A] \cdot \mathbf{u1})$$

$$/(v2 \cdot [A] \cdot u1 * v1 \cdot [A] \cdot u2 - v2 \cdot [A] \cdot u2 * v1 \cdot [A] \cdot u1).$$

then for
$$t1 * v1 \cdot [A] \cdot u1 + t2 * v2 \cdot [A] \cdot u1 = q * unull \cdot u1$$
$$t1 * v1 \cdot [A] \cdot u2 + t2 * v2 \cdot [A] \cdot u2 = q * unull \cdot u2$$
$$t1 * v1 \cdot [A] \cdot u3 + t2 * v2 \cdot [A] \cdot u3 = q * unull \cdot u3.$$
$$t1 = (b2 * v2 \cdot [A] \cdot u1 - b1 * v2 \cdot [A] \cdot u2)$$
$$/(v1 \cdot [A] \cdot u2 * v2 \cdot [A] \cdot u1 - v1 \cdot [A] \cdot u1 * v2 \cdot [A] \cdot u2).$$
$$t2 = (b1 * v1 \cdot [A] \cdot u2 - b2 * v1 \cdot [A] \cdot u1)$$
$$/(v2 \cdot [A] \cdot u1 * v1 \cdot [A] \cdot u2 - v2 \cdot [A] \cdot u2 * v1 \cdot [A] \cdot u1).$$

and the solution
$$x = ((b2 * v2 \cdot [A] \cdot u1 - b1 * v2 \cdot [A] \cdot u2))$$
$$/(v2 \cdot [A] \cdot u1 * v1 \cdot [A] \cdot u2 - v2 \cdot [A] \cdot u2 * v1 \cdot [A] \cdot u1))$$
$$* v1$$
$$+ ((b1 * v1 \cdot [A] \cdot u2 - b2 * v1 \cdot [A] \cdot u1))$$
$$/(v2 \cdot [A] \cdot u1 * v1 \cdot [A] \cdot u2 - v2 \cdot [A] \cdot u2 * v1 \cdot [A] \cdot u1))$$
$$* v2$$
$$+ t3 * unull).$$

Einstein, still incredulous: "Do an example!"

Breton: "All right. As before let
$$A = u1 * u1 + u1 * u2 + u1 * u3$$
$$+ 3 * u2 * u1 + 2 * u2 * u2 + u2 * u3$$
$$+ 3 * u3 * u1 + 3 * u3 * u2 + 3 * u3 * u3$$

The null subset of **A** is $\{t3 * (3 * u1 - u3)\}$.
Let the probe vectors
$$v1 = u1$$
$$v2 = u2.$$
Then
$$v1 \cdot [A] \cdot u1 = 1$$
$$v1 \cdot [A] \cdot u2 = 1$$
$$v1 \cdot [A] \cdot u3 = 1$$
$$v2 \cdot [A] \cdot u1 = 3$$
$$v2 \cdot [A] \cdot u2 = 2$$
$$v2 \cdot [A] \cdot u3 = 1$$
as in our earlier example.
Let
$$b = 3 * u1 - u3.$$
Then
$$unull = (3 * u1 - u3)/\sqrt{((3 * u1 - u3) \cdot (3 * u1 - u3))}$$
$$= (3 * u1 - u3)/\sqrt{10}$$
$$b \cdot unull = (3 * u1 - u3) \cdot (3 * u1 - u3)/\sqrt{10}$$
$$= \sqrt{10}$$
$$= q$$
$$b \cdot u1 = 3 = b1$$
$$b \cdot u2 = 0 = b2$$
$$b \cdot u3 = -1 = b3$$
$$x = ((-3 * 2)/(3 * 1 - 2 * 1)) * u1$$
$$+ ((3 * 1(3 * 1 - 2 * 1)) * u2$$
$$+ t3 * unull$$
$$= -6 * u1 + 3 * u2 + t3 * unull$$

which verifies as
(− 6*u1 + 3*u2 + t3*unull))
· [u1*u1+u1*u2+u1*u3
  +3*u2*u1+2*u2*u2+1*u2*u3
  +3*u3*u1+3*u2*u3+3*u3*u3]
= − 6*u1+ 9*u1- 6*u2+ 6*u2 - 6*u3 + 3*u3
= 3*u1 - 3*u3.

Einstein: "So you have not produced a solution!"

Breton: "What is wrong? Einstein is right. We have the correct solution for both the **u1** and **u2** directions, but not for the **u3** direction."

Newton, reflecting: "Is **b** =3*u1-u3 in the plane defined by the probe vectors?"

Einstein: "It is in **N(A)**!"

Newton: "Then **b·A = 0**. In fact, except for **0**, no element of **N(A)** is in the image of **A**."

Breton: "Well said, Newton. That is my error. The vector **b** needs to be an element of the image, not of the solutions. We know, for example that for

$$x \cdot A = 0$$

the *solutions* for **x** are any vector in **N(A)**. The probe vectors must be chosen appropriately. Similarly, in our present example, the probe vectors have been chosen (unconsciously) for **b** = 3*u1 - 3*u3.
  So once more I am reminded of peaches and apricots and how necessary it is to state a problem correctly."

Newton: "We used the first two equations for the problem. Do we have any doubts that using the first and third equations or the second and third equations would produce similar results?"

Now Einstein sat conspicuously silent, reflecting on the integrity of his illustrious ancestor for examining his assumptions critically.

Breton: "Let's review the trail we have followed from our present elevation. The transformation **A** transforms **V3** into a different set of vectors. It may be helpful to distinguish between the input to the transformation and the output.
  A:**V3** (input)→**V3** (output)
Because A is a rank 2 matrix
  A:**N(A)**→**0**
  For any vector **vi** not in **N(A)**
  A:{**vi**+t*unull}→ **vo**
that is **A** transforms an input line of vectors parallel to **N(A)** into a single vector in the output.
  For a line of vectors {t***vi**|t in Q} not parallel to **N(A)**
  A:{t***vi**|t in Q}→ {t***vo**|t in Q}

For an input plane of vectors
{t1•vi1 + t2•vi2|t1 and t2 in Q, vi1 or vi2 in N(A)}
      A:{t1•vi1 + t2•vi2|t1 and t2 in Q, vi1 or vi2 in N(A)}
            → {t•vo1 + vo2|t in Q}
that is, such an input plane is transformed into an output line of vectors possibly offset from the origin.
   For an input plane of vectors
{t1•vi1 + t2•vi2|t1 and t2 in Q, neither vi1 nor vi2 in N(A)}
      A:{t1•vi1 + t2•vi2|t1 and t2 in Q, neither vi1 nor vi2 in N(A)}
            → {t1•vo1 + t2•vo2|t1 and t2 in Q}
that is, such an input plane is transformed into an output plane of vectors.

Newton: "So, what is the effect of choosing the probe vectors?

Breton: "The probe vectors are input vectors to the transformation. The set {t1•v1 + t2•v2 + t3•unull} equals **V3**, and so includes all possible inputs to the transformation. The output of the transformation, however, are planes in **V3**, each containing the origin.

Newton: "So the input must be constructed to include solutions to our problem."

Breton: "Yes. The output plane must contain **b**. So the probe vectors must be so chosen that the plane
            {t1•v1 • [A] + t2•v2 • [A]|t1,t2 in Q}
contains **b**. If not, then the solution is not valid."

Einstein: "Our earlier examples, then, just happened to choose the right probe vectors."

Breton: "By chance. In the last example which did not verify
         t1•u1 • [A] =t1•(u1+u2+u3)
         t2•u1 • [A] =t2•(3•u1+2•u2+u3).
There are no *t*'s which makes their sum (points in the plane) equal to **b** =3•u1−u3."

Einstein, triumphantly: "So I was right to question your earlier reasoning!"

Breton: "Remember peaches and apricots. We need to be very careful about beginnings. There is still another aspect we can discuss."

Breton, resolutely continuing: "The entire set of solutions can be expressed as
            {**x**} = **xm** +{q1•unull}
where **xm** is orthogonal to N(A). The solution **xm** would be the one with minimum magnitude. Let us find **xm**."

Einstein, rebounding: "Always these unproven statements. Prove your assertion!"

Breton: "First notice we have discovered not merely a single solution, but an infinite set of solutions. That is, if **x** is a solution, then so is **x** + q•**unull** for any q in Q. This subset of solutions is a line of vectors parallel to the null set of **A** but offset by **x**.

Secondly, as you have noticed, Einstein, our results are dependent on the probing vectors, **v1** and **v2**. Let me show initially that if **v1** and **v2** yield **x** for a solution, then n•**v1** and n•**v2**, for any n in Q, also yield the same **x** for a solution. We have shown

x = (b2∗**v2**•[A]•**u1**−b1∗**v2**•[A]•**u2**)∗**v1**
     /(**v1**•[A]•**u2**∗**v2**•[A]•**u1** −**v1**•[A]•**u1**∗**v2**•[A]•**u2**)
   +(b1∗**v1**•[A]•**u2**−b2∗**v1**•[A]•**u1**)∗**v2**
     /(**v1**•[A]•**u2**∗**v2**•[A]•**u1** −**v1**•[A]•**u1**∗**v2**•[A]•**u2**)
   +t3•**un**).

is a solution.

For n•**v1** and n•**v2** the solution for these new probing vectors is

x = (b2∗(n•**v2**•[A]•**u1**)−b1∗(n•**v2**•[A]•**u2**)∗n•**v1**
     /(n•**v1**•[A]•**u2**∗n•**v2**•[A]•**u1** −n•**v1**•[A]•**u1**∗n•**v2**•[A]•**u2**)
   +(b1∗(n•**v1**•[A]•**u2**)−b2∗(n•**v1**•[A]•**u1**)∗n•**v2**
     /(n•**v1**•[A]•**u2**∗n•**v2**•[A]•**u1** −n•**v1**•[A]•**u1**∗n•**v2**•[A]•**u2**)
   +t3•**unull**).

Notice that all the n's cancel, leaving the solution unchanged. Consequently for a matrix of rank2, probing vectors expanded by the same scalar multiplier, all yield the same solution to **x**•[A] = **b**. As a consequence we can take one of the probes as a direction."

Einstein: "Curious."

Breton, doggedly: " First let us consider −**v1**.
For −**v1** and **v2** the solution for these new probing vectors is

x = (b2∗(**v2**•[A]•**u1**)−b1∗(**v2**•[A]•**u2**)∗(−**v1**)
     /(−**v1**•[A]•**u2**∗**v2**•[A]•**u1** +**v1**•[A]•**u1**∗**v2**•[A]•**u2**)
   +(b1∗(−**v1**•[A]•**u2**)−b2∗(−**v1**•[A]•**u1**)∗**v2**
     /(−**v1**•[A]•**u2**∗**v2**•[A]•**u1** +**v1**•[A]•**u1**∗**v2**•[A]•**u2**)
 = −(b2∗(**v2**•[A]•**u1**)−b1∗(**v2**•[A]•**u2**)∗**v1**
     /−(**v1**•[A]•**u2**∗**v2**•[A]•**u1** − **v1**•[A]•**u1**∗**v2**•[A]•**u2**)
   −(b1∗(**v1**•[A]•**u2**)−b2∗(**v1**•[A]•**u1**)∗**v2**
     /−(**v1**•[A]•**u2**∗**v2**•[A]•**u1** − **v1**•[A]•**u1**∗**v2**•[A]•**u2**)

Notice that the numerator and denominator of both summands are negative and so the negatives cancel each other. The result then for the probes −**v1** and **v2** is just

x = (b2∗(**v2**•[A]•**u1**)−b1∗(**v2**•[A]•**u2**)∗**v1**
     /(**v1**•[A]•**u2**∗**v2**•[A]•**u1** − **v1**•[A]•**u1**∗**v2**•[A]•**u2**)
   +(b1∗(**v1**•[A]•**u2**)−b2∗(**v1**•[A]•**u1**)∗**v2**
     /(**v1**•[A]•**u2**∗**v2**•[A]•**u1** − **v1**•[A]•**u1**∗**v2**•[A]•**u2**)

which is just the solution for +**v1 and v2**."

Newton: "Suppose the probe vectors are n•**v1** and **v2**

Breton: "Then
x = (b2∗(**v2**•[A]•**u1**)−b1∗(**v2**•[A]•**u2**)∗(n•**v1**)
     /(n•**v1**•[A]•**u2**∗**v2**•[A]•**u1** − n•**v1**•[A]•**u1**∗**v2**•[A]•**u2**)

$$+(b1*(n\bullet v1\cdot[A]\cdot u2)-b2*(n\bullet v1\cdot[A]\cdot u1)\bullet v2$$
$$/(n\bullet v1\cdot[A]\cdot u2*v2\cdot[A]\cdot u1 - n\bullet v1\cdot[A]\cdot u1\bullet v2\cdot[A]\cdot u2)$$
$$= (b2*(v2\cdot[A]\cdot u1)-b1*(v2\cdot[A]\cdot u2)*(n\bullet v1)$$
$$/(n*(v1\cdot[A]\cdot u2*v2\cdot[A]\cdot u1 - v1\cdot[A]\cdot u1\bullet v2\cdot[A]\cdot u2))$$
$$+(n*(b1\bullet v1\cdot[A]\cdot u2)-b2\bullet v1\cdot[A]\cdot u1)\bullet v2)$$
$$/(n*(v1\cdot[A]\cdot u2*v2\cdot[A]\cdot u1 - v1\cdot[A]\cdot u1\bullet v2\cdot[A]\cdot u2))$$
$$= (b2*(v2\cdot[A]\cdot u1)-b1*(v2\cdot[A]\cdot u2)\bullet v1$$
$$/(v1\cdot[A]\cdot u2*v2\cdot[A]\cdot u1 - v1\cdot[A]\cdot u1\bullet v2\cdot[A]\cdot u2)$$
$$+ (b1*(v1\cdot[A]\cdot u2)-b2*(v1\cdot[A]\cdot u1)\bullet v2$$
$$/(v1\cdot[A]\cdot u2*v2\cdot[A]\cdot u1 - v1\cdot[A]\cdot u1\bullet v2\cdot[A]\cdot u2)$$

which is just the solution for **v1** and **v2**."

Einstein, superciliously: "If n = 0, the denominator equals 0; so this is inadmissable."

Breton: "Thank you for the correction, Einstein. If either of the probe vectors equals **0**, our implied requirement for non-zero probe vectors would be violated. Neither is **v1**=**v2** allowed. Moreover, if both probe vectors equal **0**, a valid solution would exist only for **b** = **0**.
  I invite you to prove that interchanging the probe vectors also yields the same solution."

Einstein: "If the probe vectors are interchanged
$$\mathbf{x} = (b2*(v1\cdot[A]\cdot u1)-b1*(v1\cdot[A]\cdot u2)\bullet v2$$
$$/(v2\cdot[A]\cdot u2*v1\cdot[A]\cdot u1 - v2\cdot[A]\cdot u1\bullet v1\cdot[A]\cdot u2)$$
$$+ (b1*(v2\cdot[A]\cdot u2)-b2*(v1\cdot[A]\cdot u1)\bullet v1$$
$$/(v2\cdot[A]\cdot u2*v1\cdot[A]\cdot u1 - v2\cdot[A]\cdot u1\bullet v1\cdot[A]\cdot u2)$$
$$= -(b2*(v2\cdot[A]\cdot u1)-b1*(v2\cdot[A]\cdot u2)\bullet v1$$
$$/-(v1\cdot[A]\cdot u2*v2\cdot[A]\cdot u1 - v1\cdot[A]\cdot u1\bullet v2\cdot[A]\cdot u2)$$
$$-(b1*(v1\cdot[A]\cdot u2)-b2*(v1\cdot[A]\cdot u1)\bullet v2$$
$$/-(v1\cdot[A]\cdot u2*v2\cdot[A]\cdot u1 - v1\cdot[A]\cdot u1\bullet v2\cdot[A]\cdot u2)$$

which after canceling the minus signs yields the same solution."

Newton: "What a curiosity! Provided neither probe vector equals **0**, valid probe vectors

  v1 and **v2**
  n∗**v1** and n∗**v2**
  n∗**v1** and **v2**
  v1 and n∗**v2**

all provide the same solution."

Breton, concluding: " So we can simplify our search by letting **v1** and **v2** be directions."

Einstein: "As *I* just said. So what is the minimal solution?"

Breton, noticing, but letting the claim go unnoticed: "Let **x** be a solution and so a vector in the line of solutions parallel to **N(A)**. Let **unull** be the direction of of **N(A)**. Then
$$\mathbf{x}\cdot\mathbf{unull} = q1$$
for some q1 in Q. Now let
$$\mathbf{xm} = \mathbf{x} - q1*\mathbf{unull}.$$
Then

$$\mathbf{xm} \cdot \mathbf{unull} = 0$$

since
$$\mathbf{xm} \cdot \mathbf{unull} = \mathbf{x} \cdot \mathbf{unull} - q1*\mathbf{unull} \cdot \mathbf{unull} = q1 - q1 = 0$$
and
$$\mathbf{xm} \cdot \mathbf{xm} \le \mathbf{x} \cdot \mathbf{x} = (\mathbf{xm} + q1*\mathbf{unull}) \cdot (\mathbf{xm} + q1*\mathbf{unull})$$
$$= \mathbf{xm} \cdot \mathbf{xm} + q1^2.$$

So **xm** has a magnitude less than any other solution. In the line of solutions, a minimum exists and the entire set of solutions can be expressed as
$$\{\mathbf{x}\} = \mathbf{xm} + \{q1*\mathbf{unull}\}.$$

Einstein: "Do an example."

Breton: "Let's use a previous example.
$$A = u1*u1+u1*u2+u1*u3$$
$$+3*u2*u1+2*u2*u2+u2*u3$$
$$+3*u3*u1+3*u3*u2+3*u3*u3$$
unull $= (3*u1-u3)/\sqrt{10}$.
b $= u1+u2+u3$.
x $= u1$
Then
$$\mathbf{x} \cdot \mathbf{unull} = u1 \cdot (3*u1-u3)/\sqrt{10} = 3/\sqrt{10} = q1$$
$$\mathbf{xm} = \mathbf{x} - q1*\mathbf{unull} = u1 - q1*\mathbf{unull}$$
$$= u1 - 3*(3*u1-u3)/10$$
$$= (u1 + 3*u3)/10$$

Einstein: "Prove this is a solution."

Breton: "Simple.
$$((u1 + 3*u3)/10) \cdot [A] = (10*u1 + 10*u2 + 10*u3)/10$$
$$= u1 + u2 + u3$$
$$= b$$

Newton, meanwhile noting the results, presented to his friends this table: "Here is a table summarizing the results of our examples.

Probe vectors	Solution	Absolute Value
u1,u2	u1	1
u1 +2*u2 + u3,u2	(u1+u3)/4	1/sqrt(8)
(u1 + 3*u3)/sqrt(10),u2	(u1 + 3*u3)/10	1/sqrt(10)

Newton: "So at least for the three solutions calculated for this example what Breton claimed is true."

Breton: "It appears that finding the null set of the matrix can be an efficient preliminary to obtaining the minimum solution. Suppose then that both **unull** and some solution **x** are known. Then
$$\mathbf{xm} = \mathbf{x} - t*\mathbf{unull}$$
for some t. Further
$$\mathbf{xm} \cdot \mathbf{unull} = \mathbf{x} \cdot \mathbf{unull} - t*\mathbf{unull} \cdot \mathbf{unull} = 0$$

Thus
$$t = \mathbf{x} \cdot \mathbf{unull} / \mathbf{unull} \cdot \mathbf{unull}$$
So
$$\mathbf{xm} = \mathbf{x} - \mathbf{x} \cdot \mathbf{unull} \bullet \mathbf{unull}/\mathbf{unull} \cdot \mathbf{unull}$$
$$= \mathbf{x} \cdot [\mathbf{I} - \mathbf{unull} \bullet \mathbf{unull}/\mathbf{unull} \cdot \mathbf{unull}]$$
where **I** is the identity matrix."

Einstein, somewhat overcome, redirects the conversation: "How about rank 1 matrices."

Breton, obliging: "Suppose now a matrix **A** of rank 1. Then **N(A)** is a plane of vectors.

Einstein, taking charge: "For such a matrix, I ask again: 'What is the solution for **x** in the equation
$$\mathbf{x} \cdot \mathbf{A} = \mathbf{b}$$
where the matrix **A** and the vector **b** are given?'"

Breton: "Now let **v**, **n1**, and **n2** be three different non-planar vectors with **n1** and **n2** vectors in N(**A**). Then the vector **x** may be represented as
$$\mathbf{x} = t0 \bullet \mathbf{v} + t1 \bullet \mathbf{n1} + t2 \bullet \mathbf{n2}$$
for some t0, t1, and t2. So the solution for
$$\mathbf{x} \cdot \mathbf{A} = (t0 \bullet \mathbf{v} + t1 \bullet \mathbf{n1} + t2 \bullet \mathbf{n2}) \cdot \mathbf{A} = \mathbf{b}$$
devolves into a solution for t0 since
$$(t1 \bullet \mathbf{n1} + t2 \bullet \mathbf{n2}) \cdot [\mathbf{A}] = \mathbf{0}.$$

Newton: "What is image of **A**?"

Breton: "For a matrix of rank 1, the null set is a plane of vectors which are transformed by **A** into the single vector **0**. The entire set of vectors can be partitioned into a set of planes parallel to the null set. We can expect here that such planes will be transformed by the matrix into single vectors in its output. Now the set of solutions to a given matrix will be a plane parallel to **N(A)**. We might also note that if **x1** and **x2** are solutions, then **x1-x2** lies in the null set of **A** since
$$(\mathbf{x1} - \mathbf{x2}) \cdot \mathbf{A} = \mathbf{x1} \cdot \mathbf{A} - \mathbf{x2} \cdot \mathbf{A} = \mathbf{b} - \mathbf{b} = \mathbf{0}.$$

Einstein: "So what is a solution?"

Breton: "We can calculate as before
$$\mathbf{v} \cdot [\mathbf{A}] = (v1*a11 + v2*a21 + v3*a31) \bullet \mathbf{u1}$$
$$+ (v1*a12 + v2*a22 + v3*a32) \bullet \mathbf{u2}$$
$$+ (v1*a13 + v2*a23 + v3*a33) \bullet \mathbf{u3}$$
which we can symbolize as
$$\mathbf{v} \cdot [\mathbf{A}] \cdot \mathbf{u1} = v1*a11 + v2*a21 + v3*a31$$
$$\mathbf{v} \cdot [\mathbf{A}] \cdot \mathbf{u2} = v1*a12 + v2*a22 + v3*a32$$
$$\mathbf{v} \cdot [\mathbf{A}] \cdot \mathbf{u3} = v1*a13 + v2*a23 + v3*a33$$
Then we can write
$$\mathbf{v} \cdot [\mathbf{A}] = \mathbf{v} \cdot [\mathbf{A}] \cdot \mathbf{u1} \bullet \mathbf{u1} + \mathbf{v} \cdot [\mathbf{A}] \cdot \mathbf{u2} \bullet \mathbf{u2} + \mathbf{v} \cdot [\mathbf{A}] \cdot \mathbf{u3} \bullet \mathbf{u3}$$
so that
$$(t0 \bullet \mathbf{v} + t1 \bullet \mathbf{n1} + t2 \bullet \mathbf{n2}) \cdot \mathbf{A}$$
$$= t0 \bullet \mathbf{v} \cdot \mathbf{A} + 0$$

$$= t0*(\mathbf{v} \cdot [\mathbf{A}] \cdot \mathbf{u1} * \mathbf{u1} + \mathbf{v} \cdot [\mathbf{A}] \cdot \mathbf{u2} * \mathbf{u2} + \mathbf{v} \cdot [\mathbf{A}] \cdot \mathbf{u3} * \mathbf{u3})$$
$$= b1*\mathbf{u1} + b2*\mathbf{u2} + b3*\mathbf{u3}$$

Thus,
$$t0*\mathbf{v} \cdot [\mathbf{A}] \cdot \mathbf{u1} = b1$$
$$t0*\mathbf{v} \cdot [\mathbf{A}] \cdot \mathbf{u2} = b2$$
$$t0*\mathbf{v} \cdot [\mathbf{A}] \cdot \mathbf{u3} = b3$$

So we have three different equations which can be solved for t0. We can solve for t0 as
$$t0 = b1/\mathbf{v} \cdot [\mathbf{A}] \cdot \mathbf{u1}$$
$$t0 = b2/\mathbf{v} \cdot [\mathbf{A}] \cdot \mathbf{u2}$$
$$t0 = b3/\mathbf{v} \cdot [\mathbf{A}] \cdot \mathbf{u3}$$

So our solution becomes
$$\mathbf{x} = (t0*\mathbf{v} + t1*\mathbf{n1} + t2*\mathbf{n2})$$
$$= (b1/\mathbf{v} \cdot [\mathbf{A}] \cdot \mathbf{u1})*\mathbf{v} + t1*\mathbf{n1} + t2*\mathbf{n2}$$
or
$$\mathbf{x} = (b2/\mathbf{v} \cdot [\mathbf{A}] \cdot \mathbf{u2})*\mathbf{v} + t1*\mathbf{n1} + t2*\mathbf{n2}$$
or
$$\mathbf{x} = (b3/\mathbf{v} \cdot [\mathbf{A}] \cdot \mathbf{u3})*\mathbf{v} + t1*\mathbf{n1} + t2*\mathbf{n2}.\text{"}$$

Einstein: "Now show directly that **x** is a solution."

Breton: "All right. Let us check against the original
$\mathbf{x} \cdot \mathbf{A} = (t0*\mathbf{v} + t1*\mathbf{n1} + t2*\mathbf{n2}) \cdot \mathbf{A}$. Starting with t1 = 0 and t2 = 0 let us and first calculate
$$t0*\mathbf{v} \cdot \mathbf{A} = b1*(\mathbf{v} \cdot [\mathbf{A}] \cdot \mathbf{u1} * \mathbf{u1} + \mathbf{v} \cdot [\mathbf{A}] \cdot \mathbf{u2} * \mathbf{u2} + \mathbf{v} \cdot [\mathbf{A}] \cdot \mathbf{u3} * \mathbf{u3}) / \mathbf{v} \cdot [\mathbf{A}] \cdot \mathbf{u1}$$
$$= b1*\mathbf{u1}$$
$$+ b1*\mathbf{v} \cdot [\mathbf{A}] \cdot \mathbf{u2} * \mathbf{u2} / (\mathbf{v} \cdot [\mathbf{A}] \cdot \mathbf{u1})$$
$$+ b1*\mathbf{v} \cdot [\mathbf{A}] \cdot \mathbf{u3} * \mathbf{u3} / (\mathbf{v} \cdot [\mathbf{A}] \cdot \mathbf{u1}).\text{"}$$

Einstein: "So your **x** is not a solution."

Breton: "But it is! Did you notice,
$$b1/\mathbf{v} \cdot [\mathbf{A}] \cdot \mathbf{u1} = b2/\mathbf{v} \cdot [\mathbf{A}] \cdot \mathbf{u2} = b3/\mathbf{v} \cdot [\mathbf{A}] \cdot \mathbf{u3}$$
so
$$b1*\mathbf{v} \cdot [\mathbf{A}] \cdot \mathbf{u2} / (\mathbf{v} \cdot [\mathbf{A}] \cdot \mathbf{u1}) = b2$$
$$b1*\mathbf{v} \cdot [\mathbf{A}] \cdot \mathbf{u3} / (\mathbf{v} \cdot [\mathbf{A}] \cdot \mathbf{u1}) = b3.$$

Newton: "So the solutions Breton wrote down are indeed solutions to $\mathbf{x} \cdot \mathbf{A} = \mathbf{b}$."

Einstein: "It's too simple. Let's do an example."

Breton: "All right. Let
$$\mathbf{A} = \mathbf{u1}*\mathbf{u1} + \mathbf{u1}*\mathbf{u2} + \mathbf{u1}*\mathbf{u3}$$
$$+ 2*\mathbf{u2}*\mathbf{u1} + 2*\mathbf{u2}*\mathbf{u2} + 2*\mathbf{u2}*\mathbf{u3}$$
$$+ 3*\mathbf{u3}*\mathbf{u1} + 3*\mathbf{u3}*\mathbf{u2} + 3*\mathbf{u3}*\mathbf{u3}$$

The null subset of **A** is {t1*(2***u1** − **u2**) + t2*(3***u1** − **u3**)} for any t1 and t2 in Q.
Let
$$\mathbf{b} = \mathbf{u1} + \mathbf{u2} + \mathbf{u3}$$
Then b1 = 1 = b2 = b3.
For $\mathbf{v} = \mathbf{u1}$ we can calculate
$$\mathbf{v} \cdot [\mathbf{A}] \cdot \mathbf{u1} = v1*a11 + v2*a21 + v3*a31 = 1$$

149

$$\mathbf{v} \cdot [\mathbf{A}] \cdot \mathbf{u2} = v1*a12 + v2*a22 + v3*a32 = 1$$
$$\mathbf{v} \cdot [\mathbf{A}] \cdot \mathbf{u3} = v1*a13 + v2*a23 + v3*a33 = 1$$
Then a solution to $\mathbf{x} \cdot \mathbf{A} = \mathbf{b}$ with $t1 = t2 = 1$ is
$\mathbf{x} = (b1/\mathbf{v} \cdot [\mathbf{A}] \cdot \mathbf{u1}) \bullet \mathbf{v} + t1 \bullet \mathbf{n1} + t2 \bullet \mathbf{n2})$
  $= \mathbf{u1} + 2 \bullet \mathbf{u1} - \mathbf{u2} + 3 \bullet \mathbf{u1} - \mathbf{u3}$
  $= 6 \bullet \mathbf{u1} - \mathbf{u2} - \mathbf{u3}$
which verifies since
$$(6 \bullet \mathbf{u1} - \mathbf{u2} - \mathbf{u3}) \cdot \mathbf{A} = \mathbf{u1} + \mathbf{u1} + \mathbf{u1} = \mathbf{b}.$$
All the equations yield the same set of solutions, namely a plane of vectors parallel to the null set of **A**."

Einstein, somewhat diffidently: "Show me."

Breton: "Easily. Suppose **x1** is a solution. Then **x1** added to any vector in $\mathbf{N}(\mathbf{A})$ is also a solution. It follows that the set
$$\{\mathbf{x1} + \mathbf{N}(\mathbf{A})\}$$
is a plane of vectors parallel to $\mathbf{N}(\mathbf{A})$."

Einstein: "Are they the only solutions? Perhaps other solutions exist beyond that plane."

Breton: "No. If **x2** were any other solution, then **x1**−**x2** lies in the null set of **A** as we showed earlier.
  Now can we find the solution with the least magnitude?"

Newton, launching forward: "Again, the set of solutions can be expressed as
$$\{\mathbf{x}\} = \mathbf{xm} + \{\mathbf{y} \,|\, \mathbf{y} \text{ is a vector in } \mathbf{N}(\mathbf{A})\}$$
where **xm** is orthogonal to $\mathbf{N}(\mathbf{A})$. The solution **xm** would be the one with minimum magnitude.
  Suppose now that both **n**, a vector in the null set, and some solution **x** are known. Then
$$\mathbf{xm} = \mathbf{x} - t \bullet \mathbf{n}$$
for some t.
Further
$$\mathbf{xm} \cdot \mathbf{n} = \mathbf{x} \cdot \mathbf{n} - t \bullet \mathbf{n} \cdot \mathbf{n} = 0.$$
Thus
$$t = \mathbf{x} \cdot \mathbf{n} / \mathbf{n} \cdot \mathbf{n}.$$
So as before
$$\mathbf{xm} = \mathbf{x} - \mathbf{x} \cdot \mathbf{n} \bullet \mathbf{n} / \mathbf{n} \cdot \mathbf{n}$$
$$= \mathbf{x} \cdot [\mathbf{I} - \mathbf{n} \bullet \mathbf{n}] / \mathbf{n} \cdot \mathbf{n}$$
where **I** is the identity matrix."

Einstein, incisively: "Not so. As we have already shown, the vectors orthogonal to **n** would form a plane, not a unique vector."

Newton: "Breton where have I gone wrong?"

Breton: "Einstein is right. Why not choose *two* non-parallel vectors in $\mathbf{N}(\mathbf{A})$ and require **xm** to be orthogonal to both. Then **xm** will be unique."

Newton, accepting the suggestion: "Something like a cross product. So let **null1** and **null2** both be non-parallel vectors of **N(A)**.

Then
$$\mathbf{xm} \cdot \mathbf{null1} = 0$$
$$\mathbf{xm} \cdot \mathbf{null2} = 0$$
Also suppose a solution **x** is known. Then
$$\mathbf{x} = \mathbf{xm} + t1*\mathbf{null1} + t2*\mathbf{null2}$$
for some t1 and t2. So
$$\mathbf{xm} = \mathbf{x} - (t1*\mathbf{null1} + t2*\mathbf{null2})$$
$\mathbf{xm} \cdot \mathbf{null1} = \mathbf{x} \cdot \mathbf{null1} - (t1*\mathbf{null1} \cdot \mathbf{null1} + t2*\mathbf{null2} \cdot \mathbf{null1}) = 0$
$\mathbf{xm} \cdot \mathbf{null2} = \mathbf{x} \cdot \mathbf{null2} - (t1*\mathbf{null1} \cdot \mathbf{null2} + t2*\mathbf{null2} \cdot \mathbf{null2}) = 0$
The two unknowns, t1 and t2 may thus be solved as
$t1*\mathbf{null1} \cdot \mathbf{null1} + t2*\mathbf{null2} \cdot \mathbf{null1}) = \mathbf{x} \cdot \mathbf{null1}$
$t1*\mathbf{null1} \cdot \mathbf{null2} + t2*\mathbf{null2} \cdot \mathbf{null2}) = \mathbf{x} \cdot \mathbf{null2}$
so
$t1*\mathbf{null1} \cdot \mathbf{null1}*\mathbf{null2} \cdot \mathbf{null2} + t2*\mathbf{null2} \cdot \mathbf{null1}*\mathbf{null2} \cdot \mathbf{null2}$
$\quad = \mathbf{x} \cdot \mathbf{null1}*\mathbf{null2} \cdot \mathbf{null2}$
$t1*\mathbf{null1} \cdot \mathbf{null2}*\mathbf{null2} \cdot \mathbf{null1} + t2*\mathbf{null2} \cdot \mathbf{null2}*\mathbf{null2} \cdot \mathbf{null1}$
$\quad = \mathbf{x} \cdot \mathbf{null2}*\mathbf{null2} \cdot \mathbf{null1}$
so subtracting
$t1*\mathbf{null1} \cdot \mathbf{null1}*\mathbf{null2} \cdot \mathbf{null2} - t1*\mathbf{null1} \cdot \mathbf{null2}*\mathbf{null2} \cdot \mathbf{null1}$
$\quad = \mathbf{x} \cdot \mathbf{null1}*\mathbf{null2} \cdot \mathbf{null2} - \mathbf{x} \cdot \mathbf{null2}*\mathbf{null2} \cdot \mathbf{null1}$
that is,
$t1 = \mathbf{x} \cdot (\mathbf{null1}*\mathbf{null2} \cdot \mathbf{null2} - \mathbf{null2}*\mathbf{null2} \cdot \mathbf{null1})$
$\quad /(\mathbf{null1} \cdot \mathbf{null1}*\mathbf{null2} \cdot \mathbf{null2} - \mathbf{null1} \cdot \mathbf{null2}*\mathbf{null2} \cdot \mathbf{null1})$.
Likewise
$t2 = \mathbf{x} \cdot (\mathbf{null2}*\mathbf{null1} \cdot \mathbf{null1} - \mathbf{null1}*\mathbf{null1} \cdot \mathbf{null2})$
$\quad /(\mathbf{null2} \cdot \mathbf{null2}*\mathbf{null1} \cdot \mathbf{null1} - \mathbf{null2} \cdot \mathbf{null1}*\mathbf{null1} \cdot \mathbf{null2})$.
Consequently,
$\mathbf{xm} = \mathbf{x} - (t1*\mathbf{null1} + t2*\mathbf{null2})$
$\quad = \mathbf{x} - (\mathbf{x} \cdot (\mathbf{null1}*\mathbf{null2} \cdot \mathbf{null2} - \mathbf{null2}*\mathbf{null2} \cdot \mathbf{null1})*\mathbf{null1}$
$\quad\quad /(\mathbf{null1} \cdot \mathbf{null1}*\mathbf{null2} \cdot \mathbf{null2} - \mathbf{null1} \cdot \mathbf{null2}*\mathbf{null2} \cdot \mathbf{null1})$
$\quad\quad + \mathbf{x} \cdot (\mathbf{null2}*\mathbf{null1} \cdot \mathbf{null1} - \mathbf{null1}*\mathbf{null1} \cdot \mathbf{null2})*\mathbf{null2}$
$\quad\quad /(\mathbf{null2} \cdot \mathbf{null2}*\mathbf{null1} \cdot \mathbf{null1} - \mathbf{null2} \cdot \mathbf{null1}*\mathbf{null1} \cdot \mathbf{null2}))$
$\quad = \mathbf{x} - (\mathbf{x} \cdot (\mathbf{null1}*\mathbf{null2} \cdot \mathbf{null2}*\mathbf{null1} - \mathbf{null2}*\mathbf{null2} \cdot \mathbf{null1}*\mathbf{null1})$
$\quad\quad /(\mathbf{null1} \cdot \mathbf{null1}*\mathbf{null2} \cdot \mathbf{null2} - \mathbf{null1} \cdot \mathbf{null2}*\mathbf{null2} \cdot \mathbf{null1})$
$\quad\quad + \mathbf{x} \cdot (\mathbf{null2}*\mathbf{null1} \cdot \mathbf{null1}*\mathbf{null2} - \mathbf{null1}*\mathbf{null1} \cdot \mathbf{null2}*\mathbf{null2})$
$\quad\quad /(\mathbf{null2} \cdot \mathbf{null2}*\mathbf{null1} \cdot \mathbf{null1} - \mathbf{null2} \cdot \mathbf{null1}*\mathbf{null1} \cdot \mathbf{null2}))$
$\quad = \mathbf{x} - (\mathbf{x} \cdot (\mathbf{null1}*\mathbf{null2} \cdot \mathbf{null2}*\mathbf{null1} - \mathbf{null2}*\mathbf{null2} \cdot \mathbf{null1}*\mathbf{null1}$
$\quad\quad + \mathbf{null2}*\mathbf{null1} \cdot \mathbf{null1}*\mathbf{null2} - \mathbf{null1}*\mathbf{null1} \cdot \mathbf{null2}*\mathbf{null2})$
$\quad\quad /(\mathbf{null1} \cdot \mathbf{null1}*\mathbf{null2} \cdot \mathbf{null2} - \mathbf{null1} \cdot \mathbf{null2}*\mathbf{null2} \cdot \mathbf{null1})$
$\quad = \mathbf{x} \cdot [\mathbf{I} - (\mathbf{null1}*\mathbf{null2} \cdot \mathbf{null2}*\mathbf{null1} - \mathbf{null2}*\mathbf{null2} \cdot \mathbf{null1}*\mathbf{null1}$
$\quad\quad\quad + \mathbf{null2}*\mathbf{null1} \cdot \mathbf{null1}*\mathbf{null2} - \mathbf{null1}*\mathbf{null1} \cdot \mathbf{null2}*\mathbf{null2})$
$\quad\quad /(\mathbf{null1} \cdot \mathbf{null1}*\mathbf{null2} \cdot \mathbf{null2} - \mathbf{null1} \cdot \mathbf{null2}*\mathbf{null2} \cdot \mathbf{null1})]$
where again **I** is the identity matrix."

Breton, admiring Newton's intellectual dexterity: "If **null1** and **null2** are chosen orthogonally
$\mathbf{xm} = \mathbf{x} \cdot [\mathbf{I} - (\mathbf{null2} \cdot \mathbf{null2}*\mathbf{null1}*\mathbf{null1} + \mathbf{null1} \cdot \mathbf{null1}*\mathbf{null2}*\mathbf{null2})$
$\quad\quad /(\mathbf{null1} \cdot \mathbf{null1}*\mathbf{null2} \cdot \mathbf{null2})]$.
Newton would you construct a table of these solutions."

Newton, happily accepting the task: "Gladly.

		Solutions of $\mathbf{x} \cdot \mathbf{A} = \mathbf{b}$, $\mathbf{A}$ and $\mathbf{b}$ given	
Rank	N(A)	Solutions	xm
0	V3	0	0
1	Plane, **n1**, **n2** in N(A)	**x** = (b1/**v**•[A]•**u1**)•**v** + t1•**n1** +t2•**n2**)	**xm** = (**x** •[I − (**n1**•**n2**•**n2**•**n1** −**n2**•**n2**•**n1**•**n1** +**n2**•**n1**•**n1**•**n2** −**n1**•**n1**•**n2**•**n2**) /(**n1**•**n1**•**n2**•**n2** − **n1**•**n2**•**n2**•**n1**)]
2	Line **n** in N(A)	**x** = ((b2•**v2**•[A]•**u1** −b1•**v2**•[A]•**u2**)•**v1** +(b1•**v1**•[A]•**u2** −b2•**v1**•[A]•**u1**)•**v2**) /(**v1**•[A]•**u2**•**v2**•[A]•**u1** −**v1**•[A]•**u1**•**v2**•[A]•**u2**) +t3•**n**)	**xm** = **x** •[I − (**n**•**n**/**n**•**n**)]
3	Unique vector	**x** = **b**•((c1•(**a2**∧**a3**) +c2•(**a3**∧**a1**)+c3•(**a1**∧**a2**)) /det[A]	

Solutions for the Matrix

Einstein, somewhat pleased at his leading questions: "Now I ask:'What is the solution for the matrix **X** in the equation
$$\mathbf{v1} \cdot \mathbf{X} = \mathbf{v2}$$
where the **v1** and **v2** are given?"

Breton: "I suspect rarely does a unique answer exist. Let us start with some definitions.
$$\mathbf{X} = \mathbf{u1}\bullet\mathbf{x1}+\mathbf{u2}\bullet\mathbf{x2}+\mathbf{u3}\bullet\mathbf{x3}$$
$$\mathbf{x1} = x11\bullet\mathbf{u1}+x12\bullet\mathbf{u2}+x13\bullet\mathbf{u3}$$
$$\mathbf{x2} = x21\bullet\mathbf{u1}+x22\bullet\mathbf{u2}+x23\bullet\mathbf{u3}$$
$$\mathbf{x3} = x31\bullet\mathbf{u1}+x32\bullet\mathbf{u2}+x33\bullet\mathbf{u3}$$
$$\mathbf{v1} = v11\bullet\mathbf{u1}+v12\bullet\mathbf{u2}+v13\bullet\mathbf{u3}$$
$$\mathbf{v2} = v21\bullet\mathbf{u1}+v22\bullet\mathbf{u2}+v23\bullet\mathbf{u3}$$
$$\mathbf{v1}\cdot\mathbf{X} = (v11*x11+v12*x21+v13*x31)\bullet\mathbf{u1}$$
$$+ (v11*x12+v12*x22+v13*x32)\bullet\mathbf{u2}$$
$$+ (v11*x13+v12*x23+v13*x33)\bullet\mathbf{u3}$$

From these definitions, you can see the solution calls for determining nine unknowns, the xij, from three equations,
$$\mathbf{v1}\cdot(x11\bullet\mathbf{u1}\bullet\mathbf{u1}+x21\bullet\mathbf{u2}\bullet\mathbf{u1}+x31\bullet\mathbf{u3}\bullet\mathbf{u1}) = v21\bullet\mathbf{u1}$$
$$\mathbf{v1}\cdot(x12\bullet\mathbf{u1}\bullet\mathbf{u2}+x22\bullet\mathbf{u2}\bullet\mathbf{u2}+x32\bullet\mathbf{u3}\bullet\mathbf{u2}) = v21\bullet\mathbf{u2}$$
$$\mathbf{v1}\cdot(x13\bullet\mathbf{u1}\bullet\mathbf{u3}+x23\bullet\mathbf{u2}\bullet\mathbf{u3}+x33\bullet\mathbf{u3}\bullet\mathbf{u3}) = v21\bullet\mathbf{u3}$$
that is,
$$\mathbf{v1}\cdot(x11\bullet\mathbf{u1}+x21\bullet\mathbf{u2}+x31\bullet\mathbf{u3}) = v21$$
$$\mathbf{v1}\cdot(x12\bullet\mathbf{u1}+x22\bullet\mathbf{u2}+x32\bullet\mathbf{u3}) = v22$$
$$\mathbf{v1}\cdot(x13\bullet\mathbf{u1}+x23\bullet\mathbf{u2}+x33\bullet\mathbf{u3}) = v23.$$

Newton: "We know the solutions to these equations. Remember the solution to $\mathbf{x}\bullet\mathbf{v1} = q1$? These equations have the same form. So
$$(x11\bullet u1+x21\bullet u2+x31\bullet u3) = v21*\mathbf{qd(v1)}$$
$$(x12\bullet u1+x22\bullet u2+x32\bullet u3) = v22*\mathbf{qd(v1)}$$
$$(x13\bullet u1+x23\bullet u2+x33\bullet u3) = v23*\mathbf{qd(v1)}.$$

Breton: "Those are the orthogonal solutions. For each orthogonal solution an infinite plane of other solutions exists, as we have seen. The three planes not need not intersect, may intersect in parallel lines, or even in a point.

Einstein: "We are facing a thicket. Before trying to cut through, we can find some solutions for particular cases. For instance, if $\mathbf{v1}=\mathbf{0}$, then $\mathbf{v2}=\mathbf{0}$ also, so that $\mathbf{X}$ can be any matrix whatsoever. Similarly, if $[\mathbf{X}]=[\mathbf{0}]$, $\mathbf{v2}=\mathbf{0}$ also, for any vector $\mathbf{v1}$ whatsoever."

Newton: "But if $\mathbf{v2}=\mathbf{0}$, then $\mathbf{v1}$ need not equal zero, nor need $[\mathbf{X}]=[\mathbf{0}]$."

Breton: "So the trivial solution for a matrix of rank 0 is known. Shall we try for solutions for a matrix of rank 3?"

Newton: "Which is to say that the matrix has an inverse."

Breton: "Right. If $\mathbf{X}$ has an inverse, then
$$\mathbf{X}^{-1} = \mathbf{x2}\wedge\mathbf{x3}\bullet u1 + \mathbf{x3}\wedge\mathbf{x1}\bullet u2 + \mathbf{x1}\wedge\mathbf{x2}\bullet u3/\det(\mathbf{X})$$
and
$$\mathbf{v1} = \mathbf{v2}\cdot\mathbf{X}^{-1}.$$

Einstein: "what is $\det(\mathbf{X})$? Is it equal to $\det(\mathbf{X}^{-1})$?"

Breton: "We will do well to proceed in our discussion with formal definitions and proofs. So let me restate some definitions.

Definitions (of a matrix of rank 3)
Given
  A matrix $\mathbf{X}$ of rank 3
$$X = u1*x1+u2*x2+u3*x3$$
then
$$\det(\mathbf{X}) \equiv \mathbf{x1}\cdot(\mathbf{x2}\wedge\mathbf{x3})$$
$$\mathbf{X}^{-1} \equiv \mathbf{x2}\wedge\mathbf{x3}\bullet u1 + \mathbf{x3}\wedge\mathbf{x1}\bullet u2 + \mathbf{x1}\wedge\mathbf{x2}\bullet u3/\det(\mathbf{X})$$
end of definition

Theorem (of determinants)
  Given
      $X = u1*x1+u2*x2+u3*x3$, a matrix of rank 3;
      $\det(\mathbf{X})$;
      $\mathbf{X}^{-1}$;
      $I = u1\bullet u1+u2\bullet u2+u3\bullet u3$, the identity matrix;
      q an element of Q;

then
$$\mathbf{X} \cdot \mathbf{X}^{-1} = \mathbf{I};$$
$$\det[\mathbf{I}] = 1;$$
$$\det[q \bullet \mathbf{X}] = q^3 \bullet \det[\mathbf{X}];$$
$$\det[\mathbf{X}^{-1}] \neq 0;$$
$$\det[\mathbf{X}] = \det[\mathbf{T}[[\mathbf{X}]]$$
$$\det[\mathbf{X}] \bullet \det[\mathbf{X}^{-1}] = 1.$$

Proof:
For the first proposition:
$\mathbf{X} \cdot \mathbf{X}^{-1} = (\mathbf{u1} \bullet \mathbf{x1} + \mathbf{u2} \bullet \mathbf{x2} + \mathbf{u3} \bullet \mathbf{x3})$
$\qquad \cdot (\mathbf{x2} \wedge \mathbf{x3} \bullet \mathbf{u1} + \mathbf{x3} \wedge \mathbf{x1} \bullet \mathbf{u2} + \mathbf{x1} \wedge \mathbf{x2} \bullet \mathbf{u3})/\det(\mathbf{X})$
$= (\mathbf{u1} \bullet (\mathbf{x1} \cdot \mathbf{x2} \wedge \mathbf{x3}) \bullet \mathbf{u1}$
$\qquad + (\mathbf{u1} \bullet (\mathbf{x1} \cdot \mathbf{x1} \wedge \mathbf{x2}) \bullet \mathbf{u2} + (\mathbf{u1} \bullet (\mathbf{x1} \cdot \mathbf{x1} \wedge \mathbf{x2}) \bullet \mathbf{u3}$
$\quad + \mathbf{u2} \bullet \mathbf{x2} \cdot \mathbf{x2} \wedge \mathbf{x3}) \bullet \mathbf{u1}$
$\qquad + \mathbf{u2} \bullet \mathbf{x2} \cdot \mathbf{x3} \wedge \mathbf{x1}) \bullet \mathbf{u2} + \mathbf{u2} \bullet \mathbf{x2} \cdot \mathbf{x1} \wedge \mathbf{x2}) \bullet \mathbf{u3}$
$\quad + \mathbf{u3} \bullet \mathbf{x3} \cdot \mathbf{x2} \wedge \mathbf{x3}) \bullet \mathbf{u1}$
$\qquad + \mathbf{u3} \bullet \mathbf{x3} \cdot \mathbf{x3} \wedge \mathbf{x1}) \bullet \mathbf{u2} + \mathbf{u3} \bullet \mathbf{x3} \cdot \mathbf{x1} \wedge \mathbf{x2}) \bullet \mathbf{u3})/\det(\mathbf{X})$
$= (\mathbf{u1} \bullet (\mathbf{x1} \cdot \mathbf{x2} \wedge \mathbf{x3}) \bullet \mathbf{u1}$
$\qquad + \mathbf{u2} \bullet \mathbf{x2} \cdot \mathbf{x3} \wedge \mathbf{x1}) \bullet \mathbf{u2}$
$\qquad + \mathbf{u3} \bullet \mathbf{x3} \cdot \mathbf{x1} \wedge \mathbf{x2}) \bullet \mathbf{u3})/\det(\mathbf{X})$
$= (\mathbf{u1} \bullet \mathbf{u1} \bullet (\mathbf{x1} \cdot \mathbf{x2} \wedge \mathbf{x3})$
$\qquad + \mathbf{u2} \bullet \mathbf{u2} \bullet (\mathbf{x1} \cdot \mathbf{x2} \wedge \mathbf{x3})$
$\qquad + \mathbf{u3} \bullet \mathbf{u3} \bullet (\mathbf{x1} \cdot \mathbf{x2} \wedge \mathbf{x3}))/(\mathbf{x1} \cdot \mathbf{x2} \wedge \mathbf{x3})$
$= (\mathbf{u1} \bullet \mathbf{u1} + \mathbf{u2} \bullet \mathbf{u2} + \mathbf{u3} \bullet \mathbf{u3}$
$= \mathbf{I}$

This proves the first proposition.
For the second proposition.
$\det(\mathbf{I}) = (\mathbf{u1} \cdot \mathbf{u2} \wedge \mathbf{u3})$
$\qquad = (\mathbf{u1} \cdot \mathbf{u1})$
$\qquad = 1$

This proves the second proposition.
For the third proposition.
$q \bullet \mathbf{X} = q \bullet (\mathbf{u1} \bullet \mathbf{x1} + \mathbf{u2} \bullet \mathbf{x2} + \mathbf{u3} \bullet \mathbf{x3})$
$\det(q \bullet \mathbf{X}) = q \bullet \mathbf{x1} \cdot (q \bullet \mathbf{x2} \wedge q \bullet \mathbf{x3})$
$\qquad \qquad = q^3 \bullet (\mathbf{x1} \cdot (\mathbf{x2} \wedge \mathbf{x3}))$
$\qquad \qquad = q^3 \bullet \det(\mathbf{X})$

This proves the third proposition.
For the fourth proposition.
$\det(\mathbf{X}) = \mathbf{x1} \cdot (\mathbf{x2} \wedge \mathbf{x3})$
$\qquad = x11*x22*x33$
$\qquad \; +x12*x23*x31$
$\qquad \; +x13*x21*x32$
$\qquad \; -x11*x23*x32$
$\qquad \; -x12*x21*x33$
$\qquad \; -x13*x22*x31.$
$\qquad = x11*x22*x33$
$\qquad \; +x21*x13*x32$
$\qquad \; +x31*x12*x23$
$\qquad \; -x11*x23*x32$
$\qquad \; -x21*x12*x33$
$\qquad \; -x31*x13*x22.$
$\qquad = \det(\mathbf{T}[(\mathbf{X})]$

This proves the fourth proposition.

For the fifth proposition.
$\det(\mathbf{X}) = \mathbf{x1} \cdot (\mathbf{x2} \wedge \mathbf{x3})$
$\det(\mathbf{X}^{-1}) = \det(\mathbf{T}[(\mathbf{X}^{-1})]$
$\quad = (\mathbf{x2} \wedge \mathbf{x3}) \cdot ((\mathbf{x3} \wedge \mathbf{x1}) \wedge (\mathbf{x1} \wedge \mathbf{x2}))/(\det(\mathbf{X}))^3$
$\det(\mathbf{X}) * \det(\mathbf{X}^{-1}) = \mathbf{x1} \cdot (\mathbf{x2} \wedge \mathbf{x3}) * (\mathbf{x2} \wedge \mathbf{x3}) \cdot ((\mathbf{x3} \wedge \mathbf{x1}) \wedge (\mathbf{x1} \wedge \mathbf{x2}))$
$\quad = \mathbf{x1} \cdot (\mathbf{x2} \wedge \mathbf{x3}) * (\mathbf{x2} \wedge \mathbf{x3}) \cdot ((\mathbf{x3} \wedge \mathbf{x1}) \wedge (\mathbf{x1} \wedge \mathbf{x2}))/(\det(\mathbf{X}))^3$
Now
$(\mathbf{x3} \wedge \mathbf{x1}) \wedge (\mathbf{x1} \wedge \mathbf{x2}) = (\mathbf{x2} \cdot (\mathbf{x3} \wedge \mathbf{x1})) * \mathbf{x1} - (\mathbf{x1} \cdot (\mathbf{x3} \wedge \mathbf{x1})) * \mathbf{x2}$
$\quad = (\mathbf{x1} \cdot (\mathbf{x2} \wedge \mathbf{x3})) * \mathbf{x1} - (\mathbf{x3} \cdot (\mathbf{x1} \wedge \mathbf{x1})) * \mathbf{x2}$
$\quad = (\mathbf{x1} \cdot (\mathbf{x2} \wedge \mathbf{x3})) * \mathbf{x1}$
Therefor
$\det(\mathbf{X}) * \det(\mathbf{X}^{-1})$
$\quad = \mathbf{x1} \cdot (\mathbf{x2} \wedge \mathbf{x3}) * (\mathbf{x2} \wedge \mathbf{x3}) \cdot (\mathbf{x1} * \mathbf{x1} \cdot (\mathbf{x2} \wedge \mathbf{x3}))/(\det(\mathbf{X}))^3$
$\quad = \mathbf{x1} \cdot (\mathbf{x2} \wedge \mathbf{x3}) * \mathbf{x1} \cdot (\mathbf{x2} \wedge \mathbf{x3}) * \mathbf{x1} \cdot (\mathbf{x2} \wedge \mathbf{x3}))/(\det(\mathbf{X}))^3$
$\quad = (\mathbf{x1} \cdot (\mathbf{x2} \wedge \mathbf{x3}))^3/(\det(\mathbf{X}))^3$
$\quad = 1$
This proves the fifth proposition.

$\qquad\qquad\qquad\qquad\qquad\qquad\qquad\qquad\qquad$ qed

Breton: "If $\det(\mathbf{X}^{-1})$ is non-zero, then $\mathbf{X}^{-1}$ is also a matrix of rank3. So matrices of rank3 also have inverses of rank 3."

Newton: "This set of determinants has inverses, like quotient numbers—and like quotient vectors. Have we, in fact, defined an algebra of matrices?"

Breton: "Just so. With the inclusion of inverse matrices, the set of matrices of rank 3 do indeed constitute an algebra since given any two such matrices, **X1 and X2,**
$\qquad$ **X1** + **X2** is defined,
$\qquad$ **X1** − **X2** is defined,
$\qquad$ **X1** • **X2** is defined,
$\qquad$ **X1** / **X2** = **X1** • **X2** $^{-1}$ is defined.

Einstein: "Then we should be able to solve my question: What is the matrix **X** in the equation
$\qquad\qquad\qquad$ **v1** • **X** = **v2**
where the **v1** and **v2** are given?"

Breton: "For a matrix of rank 3, the vectors, **v1** and **v2** must obey
$\qquad\qquad\qquad$ **v1** = **v2** • **X**$^{-1}$
so let us work on these solutions first."

Einstein: "Fine, go to it."

Breton: "First, I suspect there exist more than one solution. So let us start by finding at least one solution. Suppose **X** diagonal. Then let
$\qquad\qquad$ v1 = v11•u1 + v12•u2 + v13•u3
$\qquad\qquad$ v2 = v21•u1 + v22•u2 + v23•u3
$\qquad\qquad$ X = g1•u1•u1 + g2•u2•u2 + g3•u3•u3

If we can find the g's in terms of the v's we will have a solution. So expanding $\mathbf{v1} \cdot \mathbf{X} = \mathbf{v2}$ we have

v11•u1 + v12•u2 +v13•u3
   • [g1•u1•u1 +g2•u2•u2 +g3•u3•u3]
    = v21•u1 + v22•u2 +v23•u3

that is,
v11*g1•u1 + v12*g2•u2 +v13*g3•u3
    = v21•u1 + v22•u2 +v23•u3

so
    g1 = v21/v11
    g2 = v22/v12
    g3 = v23/v13

You can see easily that
  $\mathbf{X}$= v21•u1•u1/v11 + v22•u2•u2/v12 +v23•u3•u3/v13
is a solution."

Newton: "How about the inverse."

Breton: "For the diagonal case
    $\mathbf{X}^{-1}$ = u1•u1/g1 +u2•u2/g2 +u3•u3/g3
and
    $\mathbf{v1} = \mathbf{v2} \cdot \mathbf{X}^{-1}$
expands as
v11•u1 + v12•u2 +v13•u3
 = v21•u1 + v22•u2 +v23•u3
   • u1•u1/g1 +u2•u2/g2 +u3•u3/g3.
Then for the same values of the g's
v11•u1 + v12•u2 +v13•u3
 = v21•u1 + v22•u2 +v23•u3
   • [v11•u1•u1/v21
    +v12•u2•u2/v22
    +v13•u3•u3/v23]."

Einstein: "Too easy. How about other solutions?"

Breton: "Do you trust me to find others?"

Newton: "No need for trust. Just produce what you say is a solution; we can easily verify it."

Breton: "All right. I will use a method for finding rank3 solutions which can be modified to find solutions for rank2 and rank1 matrices also.
 To find the solutions of $\mathbf{v1} \cdot \mathbf{X} = \mathbf{v2}$ of rank 3 in general, chose three non- parallel directions, **ux1**, **ux2**, and **ux3**. Next, define
    t1 ≡ $\mathbf{v2} \cdot \mathbf{ux1}/(\mathbf{v1} \cdot \mathbf{ux1})$
    t2 ≡ $\mathbf{v2} \cdot \mathbf{ux2}/(\mathbf{v1} \cdot \mathbf{ux2})$
    t3 ≡ $\mathbf{v2} \cdot \mathbf{ux3}/(\mathbf{v1} \cdot \mathbf{ux3})$.
Then
    $\mathbf{X}$= t1*ux1•u1 + t2*ux2•u2 + t3*ux3•u3
is a solution, since

$$v1 \cdot X = v1 \cdot [t1*ux1*u1 + t2*ux2*u2 + t3*ux3*u3]$$
$$= t1*v1 \cdot ux1*u1 + t2*v1 \cdot ux2*u2 + t3*v1 \cdot ux3*u3$$
$$= (v2 \cdot u1/v1 \cdot ux1)*v1 \cdot ux1*u1$$
$$+ (v2 \cdot u2/v1 \cdot ux2)*v1 \cdot ux2*u2$$
$$+ (v2 \cdot u3/v1 \cdot ux3)*v1 \cdot ux3*u3$$
$$= v2 \cdot [u1*u1 + u2*u2 + u3*u3]$$
$$= v2.$$

Newton: "That's like magic. How did you know how to define the t's?"

Breton: "You didn't trust me! Still the solution is verified."

Newton: "So an infinity of choices is available, for all the different choices of **ux**'s possible."

Einstein: "Suppose each **uxi** = **ui**."

Breton: "Then
$$X = t1*u1*u1 + t2*u2*u2 + t3*u3*u3$$
a diagonal matrix
$$X = v2 \cdot u1/(v1 \cdot u1*ux1*u1 + t2*ux2*u2 + t3*ux3*u3$$
and
$$t1 \equiv v2 \cdot u1/(v1 \cdot u1)$$
$$t2 \equiv v2 \cdot u2/(v1 \cdot u2)$$
$$t3 \equiv v2 \cdot u3/(v1 \cdot u3).$$

Newton: "Which is precisely our former solution."

Einstein, moving to recapture the initiative: "How about the rank2 solutions?"

Breton: "To find the rank 2 solutions of **v1** • **X** = **v2** chose two non-parallel directions, **ux1**, **ux2**, Next define

$$t1 \equiv v2 \cdot u1/v1 \cdot ux1$$
$$t2 \equiv v2 \cdot u2/v1 \cdot ux2$$
$$t3 \equiv v2 \cdot u3/v1 \cdot ux2$$

Then
$$X = t1*ux1*u1 + t2*ux2*u2 + t3*ux2*u3$$
is a solution, since
$$v1 \cdot X = v1 \cdot [t1*ux1*u1 + t2*ux2*u2 + t3*ux2*u3]$$
$$= t1*v1 \cdot ux1*u1 + t2*v1 \cdot ux2*u2 + t3*v1 \cdot ux2*u3$$
$$= (v2 \cdot u1*v1 \cdot ux1*u1/v1 \cdot ux1)$$
$$+ (v2 \cdot u2*v1 \cdot ux2*u2/v1 \cdot ux2)$$
$$+ (v2 \cdot u3*v1 \cdot ux2*u3/v1 \cdot ux2)$$
$$= v2 \cdot [u1*u1 + u2*u2 + u3*u3]$$
$$= v2$$

Six similar variations may be formed for any arbitrary choice of any two non-parallel directions."

Newton: "So the method for the magic is clear. Let me try the rank 1 solutions of **v1**·**X** = **v2**. Chose any direction **ux1** not orthogonal to **v1**. Next, define

$$t1 \equiv \mathbf{v2}\cdot\mathbf{u1}/\mathbf{v1}\cdot\mathbf{ux1}$$
$$t2 \equiv \mathbf{v2}\cdot\mathbf{u2}/\mathbf{v1}\cdot\mathbf{ux1}$$
$$t3 \equiv \mathbf{v2}\cdot\mathbf{u3}/\mathbf{v1}\cdot\mathbf{ux1}$$

Then
$$\mathbf{X} = t1*\mathbf{ux1}*\mathbf{u1} + t2*\mathbf{ux1}*\mathbf{u2} + t3*\mathbf{ux1}*\mathbf{u3}$$
is a solution, since
v1·**X** = **v1**·[t1∗**ux1**∗**u1** + t2∗**ux1**∗**u2** + t3∗**ux1**∗**u3**]
 = t1∗**v1**·**ux1**∗**u1** + t2∗**v1**·**ux1**∗**u2** + t3∗**v1**·**ux1**∗**u3**
 = **v2**·**u1**∗**v1**·**ux1**∗**u1**/**v1**·**ux1**
  + (**v2**·**u2**∗**v1**·**ux1**∗**u2**/**v1**·**ux1**
  + (**v2**·**u3**)∗**v1**·**ux1**∗**u3**/**v1**·**ux1**
 = **v2**·[**u1**∗**u1** + **u2**∗**u2** + **u3**∗**u3**]
 = **v2**.

Breton: "Suppose **ux1** = **uv1**."

Newton: "Then
X = t1∗**uv1**∗**u1** + t2∗**uv1**∗**u2** + t3∗∗**uv1** **u3**
 = (**v2**·**u1**)∗**u1**∗**uv1**/v1
  + (**v2**·**u2**)∗**u2**∗**uv1**/v1
  + (**v2**·**u3**)∗**u3**∗**uv1**/v1
 = **v2**·(**u1**∗**u1** + **u2**∗**u2** + **u3**∗**u3**)∗**uv1**/v1
 = **v2**∗**uv1**/v1.

Breton: "which can be rewritten
$$\mathbf{X} = = \mathbf{v2}*qd(\mathbf{v1}),$$
an outer product.
Furthermore for any **v** and a matrix **X** of rank 1 which solves the equation **v1**·**X** = **v2**
v·**X** = **v**·[t1∗**u1**∗**ux1** + t2∗**u2**∗**ux1** + t3∗**u3**∗**ux1**]
 = t1∗**v**·**ux1**∗**u1** + t2∗**v**·**ux1**∗**u2** + t3∗**v**·**ux1**∗**u3**
 = **v2**·**u1**∗**v**·**ux1**∗**u1**/**v1**·**ux1**
  + (**v2**·**u2**∗**v**·**ux1**∗**u2**/**v1**·**ux1**
  + (**v2**·**u3**∗**v**·**ux1**∗**u3**/**v1**·**ux1**
 = (**v**·**ux1**/(**v1**·**ux1**))
   (**v**·**ux1**/(**v1**·**ux1**)) + **u2**∗**u2** + **u3**∗**u3**]
 = (**v**·**ux1**/(**v1**·**ux1**))∗**v2**
Thus **X** maps all vectors into a line with a unique direction.

Transformation into Theoretical Physics

Einstein: "As you say, Breton, none of this intricate vector algebra is Theoretical Physics. Show us how our results become Theoretical Physics!"

Breton: "First of all in terms of physical units."

Newton: "We have already made a start in that transformation when we insisted that vector multiplications corresponds to areas in contrast to vectors themselves which have lengths or magnitudes. Mathematically we might have considered an expression like
$$v1 + v2 \bullet v3$$
which would make no sense for Theoretical Physics because the units could not be made to correspond."

Breton: "So in our development of the algebra we accommodated this requirement of Theoretical Physics somewhat. We have seen that for **r1** and **r2** as location vectors, areas can be defined as
$$r1 \bullet r2$$
$$r1 \wedge r2$$
$$r1 * r2$$

that is, as a scalar, a vector, or a transformation. Let us symbolize the transformational area as $\mathbf{A} = r1*r2$; the vectorial area as $\mathbf{a} = r1 \wedge r2$; the scalar area as $a = r1 \bullet r2$. These latter areas are related to the transformational area as
$$\mathbf{a} = c[\mathbf{A}]$$
and
$$a = tr[\mathbf{A}]$$
where **c** is the curl matrix operator and $tr[\mathbf{A}]$ In is the trace of **A**."

Newton: "Of course. Why didn't we recognize this earlier?"

Breton: "The scalar areas are related by
$$r1 \bullet [r1*r2] \bullet r2 = r1 \bullet [C(r1 \wedge r2)] \bullet r2 + r1 \bullet r2 * r1 \bullet r2$$
where **C** is the curl vector operator. This relationship is the vectorial expression of the Pythagorean theorem."

Einstein: "Not necessarily tied to Euclidean geometry."

Breton: "Correct. In addition to these simply local ideas, material duals are possible. For particles **x1** and **x2** Theoretical Physics conceives ideas like
$$x1*x2$$
$$r1*x2$$
$$x1*r2$$
with similar vectorial and scalar relatives as material areas and with analogous relationships between them."

Newton: "You are greatly expanding the treasury of ideas of Theoretical Physics."

Breton: "Just as we noted earlier in tp1.1. Restrictions first, then expansion. But this is only a start. Acknowledging the basic physical units L, V, and F, Theoretical Physics also addresses, in addition to combinations like L*L, others like

$$V*V$$
$$F*F$$
$$L*V$$
$$L*F$$
$$V*F$$

As you can see this is an enormous scope of ideas, whose investigation will have to wait for further development."

Newton: "What do you mean?"

Breton: "We have developed a vectorial *algebra*. In deference to your illustrious ancestor, you may be keen to investigate the development of a vectorial *calculus*. That development can be expected to reflect on our considerations of vectorial areas, no less than Isaac Newton's calculus did on scalar areas."

Newton: "Let's start now."

Breton: "Patience. Let it be the subject of a new book."

Einstein: "Areas, areas. How about volumes?"

Breton: "We have seen the scalar and vectorial triple products relate to scalar and vectorial volumes. The elements of these vectorial combinations are expanded in Theoretical Physics to combinations of basic and derived physical units. Here again a large scope of inquiry awaits our investigation which should be enhanced when we have acquired the tools of a vectorial calculus."

References

Newton: "How do references come into play? Yesterday we talked about references, but not in terms of observations. Please discuss this matter further."

Breton: "To further the transformation of mathematical ideas into Theoretical Physics, a way is needed to identify change in the observed object. So Theoretical Physics requires a starting condition as a reference for the change.
 Start by first referencing a hypothetical, material **universe at rest**--a universe in which nothing is moving."

Einstein: "What is physical about a universe at rest? Isn't Theoretical Physics supposed to develop ideas for understanding physical reality?"

Breton: "The fictitious universe at rest is merely an intellectual convenience and not strictly necessary. Any method of identifying physical observables as they move and change is acceptable. The reference to a universe at rest is convenient because it avoids the necessity of establishing initial conditions of motion."

Einstein: "Strictly your invention?"

Breton: "With no allusion that such a universe ever existed. So will you let me proceed?"

Einstein: "With that understanding, yes."

Breton: "Next, associate this hypothetical universe of physical matter with **V3**.
   There are two ways of making this association. In the first way, associate every particle (**x**) of the material universe at rest with a location, **x**, of **V3**, denoting its distance and direction referred to the observer. Of course, there may be locations **x** of **V3** for which nothing material exists to be observed. Assign to those locations the idea of **null matter** that is, a location in **V3** for which no matter exists in the universe of rest. Then each vector of **V3** is associated in the universe at rest with the location of a material object, either observable or null. Thus associated, **V3** is properly labeled **V3(x)**."

Newton: "What is this 'particle' you are talking about?"

Breton: "Don't you remember? It was one of the last items defined yesterday and can now be found in tp1.1. Let me repeat the definition.

Definition (particle)
    Given
        S(**x**) a set of objects with a topology
        x a member of S(**x**)
        VT(**x**) the set of all open sets containing **x**

    then
    A **particle** (**x**) is the intersection of all the subsets of VT(**x**)
                                             end of definition

---

As a set function, **x** identifies extended physical matter in the hypothetical universe by associating it uniquely with a set of location vectors in **V3**. With reference to this hypothetical universe, directly observable primary changes in matter can be described as functions of **x** symbolized as follows:

Observable	At Rest	Afterwards	Idea
extension	**x**	**r(x)**	location
motion	0	**v(x)**	velocity

Material References

Material/Spatial References

Einstein: "So observable objects in the universe are associated with our set of vectors, **V3**. We set an initial association as the universe at rest and call it **V3(x)**; subsequently the association is called **V3(r)**."

Breton: "Consider the function, **r**:**V3(x)**→**V3(r)**, denoted by **r(x)**, which describes the change, if any, in the location of an observable particle, (**x**). For instance, **r(x)** = **x** indicates that a given particle **x** is at its reference location.

With the assumption of null matter for logical completion, **r(x)** not only describes the location of individual particles **x**, but is also a bijective map of **V3** into and onto itself. Considered as a bijective transformation, **r(x)** will then generate an inverse transformation, **x(r)**.

We may interpret **x(r)** specifically. Fixing our attention on a particular location **r1**, **x(r1)** describes the different observable particles which occupy the location **r1**.

Thus the second way of associating the material universe with **V3** becomes apparent.

In terms of primary functions, then, Theoretical Physics proposes this schema:
  x:{extended objects and null matter} →**V3(x)**
  r:{physical location} →**V3(r)**
  r:**V3(x)** →**V3(r)**
  x:**V3(r)** →**V3(x)**

The identifying set functions, **x** or **r**, with their domains in observable universe and their ranges in **V3**, continue their role as identifiers through the functions **r(x)** and **x(r)** reflecting changes in the observable universe as transformations of **V3**."

Newton: "Why do you call them set functions?"

Breton: "They are not functions of a real variable. They connect two different sets: the universe at rest to **V3** and vice-versa."

Einstein: "Why 'real' variables? Wouldn't our quotient numbers do as well?"

Breton: "Yes, of course. I should rather have said complete rather than real numbers. In any case quotient numbers suffice."

Newton: "Continue."

Breton: "The two transformations, **r(x)** and **x(r)** correspond to two different ways of observing physical reality. We may either fix our attention on an object as it moves from one location to another, or alternatively we may fix our attention on only one location and observe what passes through it.
 The first way of observation, **r(x)**, is called the **material** reference; the second of way of observation, **x(r)**, is called the **fixed–local**, or often simply the **local** reference.
 The two references of Theoretical Physics lead to a duality of descriptions and analyses of physical reality. Any statement made in the material reference corresponds to a dual statement in the local reference, that is f(**r(x)**) = f(**x(r)**)."

Einstein: "Aren't you confusing transformations with functions?"

Breton: "Thank you again, Einstein. To distinguish the global transformation from a particular function the following symbology is used:

	Global	Local
Location	**r(x)** or **r|x**	**r(x1)** or **r|x1**
Matter	**x(r)** or **x|r**	**x(r1)** or **x|r1**

Symbols for Global or Local References

 The two related functions, **r(x,**a**)** and **x(r,**a**)** are called **primary functions** of Theoretical Physics.
 Since **x(r1)** means **x** is at **r1**, it follows that
$$r(x(r1)) = r1.$$
Dually,
$$x(r(x1)) = x1.$$

Observations

Newton: "How do these references relate to observations?"

Breton: "Physical observations are ordered, that is one observation follows another. Any observation comes after or before another. Theoretical Physics proposes numbering the ordered observations as a real variable, symbolized as *a*."

Einstein: "Why 'real' variables? Wouldn't our quotient numbers do as well?"

Breton: "Yes, of course. Again, I should rather have said 'complete' rather then 'real' numbers. In any case quotient numbers suffice."

Breton: "We are usually concerned with a local perspective at observation *a1*. Usually we take for reference the particle **x1** whose location at observation *a1* is **r1(x1**,a1) and the local or material topologies referenced to **r1** and **x1**.
   We can now express a fundamental principle distinguishing both Physics and Theoretical Physics from Mathematics.

The Principle of Non-Collocation

For the same observation,
   two different particles cannot occupy the same position.

Symbolically,

$$\{r(x2,a1) = r(x1,a1)\} \rightarrow x2 \text{ is } x1$$

and

$$\{x(r2,a1) = x(r1,a1)\} \rightarrow r2 \text{ is } r1.$$

Newton, with a sniff of his aquiline nose: "Why does the principle of non-contradiction matter?"

Breton: "As a first consequence of the principle of non-collocation, notice that in Theoretical Physics **r(x1**,a1) cannot be decomposed into

$$r1(\mathbf{x1},a1)*u1 + r2(\mathbf{x1},a1)*u2 + r3(\mathbf{x1},a1)*u3$$

but rather

$$r(\mathbf{x1},a1) = r1(\mathbf{y1},a1)*u1 + r2(\mathbf{y2},a1)*u2 + r3(\mathbf{y3},a1)*u3$$

for some **y1**, **y2**, and **y3**, because **x1** at observation *a1* occupies location **r(x1**,a1), not r1*u1."

Einstein, inquiringly: "How does this fit into vector algebra?"

Breton: "Let's consider **r(x1**,a1), the position of particle **x1** at observation a1, as the vector **v1**. In the set of vectors **v1+v1** = 2*v1. Then what can be meaning of

$$r(\mathbf{x1},a1)+r(\mathbf{x1},a1)?$$

Newton: "So is

$$r1(\mathbf{x1},a1)+r1(\mathbf{x1},a1) = 2*r1(\mathbf{x1},a1)?"$$

Einstein, challengingly: "What do you say, Breton?"

Breton, accepting the challenge: "The symbol **r1**(**x1**,a1) refers to the particle **x1** which is at position **r1** at observation *a1*. The symbol 2∗**r1**(**x1**,a1) refers to the particle **x1** which is at position 2∗**r1** at observation *a1*. Physically can the same particle be at two different places at the same time?"

Einstein: "Of course not! What a strange world it would be if we could walk right through walls!"

Breton: "As stated
$$\mathbf{r1}(\mathbf{x1},a1) + \mathbf{r1}(\mathbf{x1},a1) = 2*\mathbf{r1}(\mathbf{x1},a1)$$
cannot be allowed for the science of Physics because the proposition violates the principle of non-collocation. If the particle **x1** is at location **r1** at observation a1, it cannot simultaneously be located at 2∗**r1**."

Einstein: "So our efforts at constructing a vector algebra come to naught. We cannot apply vector algebra to Theoretical Physics!"

Breton: "Not so fast. We could have
$$\mathbf{r1}(\mathbf{x1},a1) + \mathbf{r1}(\mathbf{x1},a1) = 2*\mathbf{r1}(\mathbf{y1},a1)$$
for some particle **y1** other than **x1**."

Einstein: "Why not
$$\mathbf{r1}(\mathbf{x1},a1) + \mathbf{r1}(\mathbf{x1},a1) = 2*\mathbf{r1}(\mathbf{x1},a2)?$$"

Breton: "Possibly, but not generally. The particle **x1** may never arrive at position 2∗**r1** and even if it does, it may not arrive at observation a2. Let us rather put
$$\mathbf{r1}(\mathbf{x1},a1) + \mathbf{r1}(\mathbf{x1},a1) = 2*\mathbf{r1}(\mathbf{y2},a2).$$"

Einstein: "So the expression **r1**(**x1**,a1)+ **r1**(**x1**,a1) is an ambiguous expression for Theoretical Physics. The expression simply designates a location where neither the particle nor the observation is specified."

Breton: "Don't give up so easily. For different particles we could have generally
$$\mathbf{r1}(\mathbf{x1},a1) + \mathbf{r2}(\mathbf{x2},a1) = \mathbf{r3}(\mathbf{x3},a1)$$
with the restriction that **x1** = **x2** = **x3** is disallowed.
 Isn't this just what we learned before, that to use mathematical ideas for Theoretical Physics they must first be restricted?"

Einstein, mulling the matter over: "We might also have
$$\mathbf{r1}(\mathbf{x1},a1) + \mathbf{r2}(\mathbf{x2},a2) = \mathbf{r3}(\mathbf{x3},a3).$$"

Breton: "Thank you. I stand corrected. The restriction is simply that the principle of non-collocation must be respected. Then we see that once restricted, the sum
$$\mathbf{r1} + \mathbf{r2} = \mathbf{r3}$$
in Theoretical Physics expands into panoply of different expressions depending on the references."

Newton: "The same can be said of
$$x1 + x2 = x3.$$

Breton: "Of course, by duality."

Einstein, recalling the axioms for the set of vectors: "How about
$$v1+(v2+v3) = (v1+v2)+v3?"$$

Breton: "If the vector
$$r4(y,a4) = r1(x1,a1) + (r2(x2,a2) + r3(x3,a3))$$
then the vector
$$(r1(x1,a1) + r2(x2,a2)) + r3(x3,a3)$$
is located at **r4** which then must be the position of **y** at observation a4. So again the axiom of association holds in Theoretical Physics, provided non-collocation is respected.

Newton, rather enjoying the conversation, adds: "Let me add that vector subtraction similarly cannot be used unrestrictedly for Theoretical Physics either."

Einstein, questioning: "Then quantum mechanics is excluded from Physics as science?"

Breton: "It would be distracting to deal with quantum theories now. We shall discuss the subject anon. Perhaps it is not science but rather technology. Wasn't your illustrious ancestor involved in their initial development but later had strong misgivings. It should be interesting to examine quantum theories with the tools of Theoretical Physics which were not available to your illustrious ancestor."

Einstein: "I look forward to your thoughts."

Breton: "We might reflect on thoughts that lie beyond and above Physics. There is another way to express the schema linking the physical world to the intellectual ideas. The physical universe is its matter. By His act of creation God materialized the idea of a three dimensional set of vectors. In so doing He gives these ideas a new material existence without in the least compromising them as ideas.
God's purpose is revealed by his Messiah, Jesus, who comes to spiritualize matter, that is, to give matter a new spiritual existence. Jesus does this by revealing Himself as God's incarnation, that is God becoming part of the material universe without in the least compromising His divinity."

Einstein: "Your belief in God does lead to new vistas."

Breton: "Belief? We *know* and have *proven* that God exists by logical argument."

Newton: "Knowledge of his existence, yes, but not proof of his purposes."

Breton: "Newton, you are correct of course. This knowledge can only come from God who freely reveals it. But once revealed we can marvel how well it comports with Theoretical Physics. The duality in Theoretical Physics is a reflection in the physical world of the centrality of Jesus Christ, who revealed Himself by his life, death, and resurrection as the unique divine/human duality. As such, He is the origin, the one who continues, and the destiny of the material universe. It is in Him that the "groanings"of material creation find resolution. While the mystery of Jesus is valued especially for enlightenment of the human condition, it is not without light for the physical universe too."

Newton, dismissively: "So it does not arise from Theoretical Physics, however much it may comport with it. As such it is distracting us from our inquiry."

Breton: "Not so much a distraction as a frame or background for our reasoning, but not the picture itself."

Einstein, with a touch of impatience: "Let us get on with our task. We have seen that vectorial plus and minus can be brought into the framework of Theoretical Physics with restrictions. How about vector multiplication? Can vector multiplication be brought into Theoretical Physics?"

Vectorial Multiplication in Theoretical Physics

Breton: "We have made a good start on answering your question by noting that inner products and areas are related. Let
$$v1 = r1(x1, a1)$$
what is **r1**(**x1**,a1)•**r1**(**x1**,a1)?"

Einstein: "In Mathematics for **v1** = q1***uv1**
$$v1 \cdot v1 = q1*q1.$$
So what physical meaning can **r1**(**x1**,a1)•**r1**(**x1**,a1) have?"

Breton: "Then **r1**(**x1**,a1)•**r1**(**x1**,a1) becomes a reference scalar area. This is a derived idea of Theoretical Physics which can be compared to other derived ideas like
$$r1(x1,a1) \cdot r2(x1,a2)$$
$$r1(x1,a1) \cdot r1(x2,a2).$$

Newton: "Why not
$$x1(r1,a1) \cdot x1(r1,a1)$$
$$x1(r1,a1) \cdot x2(r1,a2)$$
$$x1(r1,a1) \cdot x1(r2,a2)?$$

Breton: "Why not indeed? These are all valid ideas of Theoretical Physics."

Einstein: "Then why not
$$r1(x1,a1) \cdot x1(r1,a1)$$
$$r1(x1,a1) \cdot x2(r1,a2)$$
$$r1(x1,a1) \cdot x1(r2,a2)$$
$$x1(r1,a1) \cdot r2(x1,a2)$$
$$x1(r1,a1) \cdot r1(x2,a2)?"$$

Breton: "Yes, of course. Again we see a single mathematical idea expanded into a panoply of related idea of Theoretical Physics."

Newton: "This same development surely applies to cross products and outer products!"

Breton: "Yes most certainly. So Theoretical Physics embraces a great many ideas with which to analyze observed traces and tracks."

Einstein: "But none of these areas are observable. I do observe, however, physical areas."

Breton: "So it might seem. Rather we observe extended objects, from whose surfaces we might calculate areas. The kind of area you refer to comes from integration of observed particles rather than from simpler multiplicands."

Newton: "So there are ideas of areas beyond the numerous ones we are now considering?"

Breton: "Yes, but all ideas of area in Theoretical Physics have units of L*L."

Newton: "Ideas having units of L*L form a vast collection of which we are only contemplating a small part?"

Breton: "Yes, only those which can be defined algebraically."

Newton: "The whole of which you faithfully promise to address anon?"

Breton: "You have my promise."

Einstein: "How about the proposition
$$v1 \cdot (v2+v3) = v1 \cdot v2 + v1 \cdot v3?$$

Breton: "Expressed in the primary variables of Theoretical Physics we might have
$$r1(x1,a1) \cdot r2(x1,a2) + r1(x1,a1) \cdot r3(x1,a3)$$
for $v1 \cdot v2 + v1 \cdot v3$
and
$$r1(x1,a1) \cdot (r2(x1,a2) + r3(x1,a3))$$
for $v1 \cdot (v2+v3)$."

Einstein: "So one area equals the sum of two other areas."

Breton: "Correct, so long as the principle of non-collocation is respected."

Einstein: "So
$$r1(x1,a1) \bullet (r2(x1,a2) + r2(x1,a2))$$
does not equal
$$r1(x1,a1) \bullet r2(x1,a2) + r1(x1,a1) \bullet r2(x1,a2)$$
because the first one might violate the principle of non-collocation."

Breton: "Just so. Have you noticed that when we come to areas the principle of non-collocation no longer applies? Whereas
$$r1(x1,a1) + r1(x1,a1)$$
is meaningless or at most ambiguous
$$r1(x1,a1) \bullet r1(x1,a1)$$
is a well described area in Theoretical Physics."

Newton: "Because we are no longer dealing with locations."

Einstein, moving the discussion: "How can $r1 \wedge r2$ be an area?"

Breton: "Not a scalar area, but a vector area."

Newton: "Another kind of area?"

Breton: "Another kind of algebraic area. Now
$$r1(x1,a1) \wedge r1(x1,a1) = 0."$$

Newton: "So does $r1(x1,a1) \wedge r1(x2,a2)$."

Einstein: "How about $-(r2(x1,a2) \wedge r1(x1,a1))$?"

Breton: "Since
$$r1(x1,a1) \wedge r2(x1,a2) = \text{abs}(r1(x1,a1)) * \text{abs}(r2(x1,a2))$$
$$* \sin(\text{angle}(r1(x1,a1),r2(x1,a2)))$$
$$* \text{un}(r1(x1,a1),r2(x1,a2))$$
then
$$-r1(x1,a1) \wedge r2(x1,a2) = -\text{abs}(r1(x1,a1)) * \text{abs}(r2(x1,a2))$$
$$* \sin(\text{angle}(r1(x1,a1),r2(x1,a2)))$$
$$* \text{un}(r1(x1,a1),r2(x1,a2)).$$

Einstein: "Then what is the location of the product?"

Breton: "The product is an area, not a location. As a vector, a vectorial area has both a magnitude and a direction."

Newton: "So all the different types of scalar areas have vector areas as counterparts."

Breton: "Yes, but remember these are ideas which may be used to describe or explain physical observations."

Einstein: "I suppose outer products follow the same development."

Breton: "Yes."

Einstein: "So **r1**(**x1**,a1)∗**r1**(**x1**,a1) is a transformational area?"

Breton: "Yes, it transforms any location, **r1**(**x1**,a1) into a vectorial volume as
$$(\mathbf{r1}(\mathbf{x1},a1) \bullet \mathbf{r1}(\mathbf{x1},a1)) \ast \mathbf{r1}(\mathbf{x1},a1).$$

Newton: "And so also for the many combinations discussed earlier."

Einstein, leading on: "How about the triple products as volumes?"

Breton: "The scalar triple product produces scalar volumes, as for example,
$$\mathbf{r1}(\mathbf{x1},a1) \bullet (\mathbf{r2}(\mathbf{x1},a2) \wedge \mathbf{r3}(\mathbf{x1},a3)).$$

Newton: "That would be a volume associated with the material trace of the particle **x1**."

Breton: "One could graph the area's growth as a function of a2 and a3."

Newton: "The starting volume could equal 0 as
$$\mathbf{r1}(\mathbf{x1},a1) \bullet (\mathbf{r1}(\mathbf{x1},a1) \wedge \mathbf{r1}(\mathbf{x1},a1)) = 0$$
and then grow along the trace of **x1**."

Einstein: "And again a large panoply of similar ideas from the variations considered earlier. But how about the vector triple product?"

Breton: "We proved
$$\mathbf{v1} \wedge (\mathbf{v2} \wedge \mathbf{v3}) = (\mathbf{v1} \bullet \mathbf{v3}) \ast \mathbf{v2} - (\mathbf{v1} \bullet \mathbf{v2}) \ast \mathbf{v3}$$
which transforms for Theoretical Physics into, for example,
r1(**x1**,a1)∧(**r2**(**x1**,a2)∧**r3**(**x1**,a3))
= (**r1**(**x1**,a1)•**r3**(**x1**,a3))∗**r2**(**x1**,a2))
− (**r1**(**x1**,a1)•**r2**(**x1**,a2))∗**r3**(**x1**,a3))
and the many variations mentioned above."

Newton: "So the one vectorial volume can be divided into two other vectorial volumes, each of which may be analyzed as functions of the observational index along a trace."

Some Reflections on the Science of Physics

Einstein, reflecting: "The number of such ideas is overwhelming!"

Breton, expansively: "Remember the ideas dealt with so far only deal with the primary variable of extension. But we can go further. The vectors need not carry only units of extension (L). We may also consider derived ideas like energy or power. For instance, we can consider units like F*L, which are the units of energy. Then
$$f(\mathbf{x1},a1) \bullet r(\mathbf{x1},a1) \text{ is scalar energy}$$
$$f(\mathbf{x1},a1) \wedge r(\mathbf{x1},a1) \text{ is vectorial energy}$$
$$f(\mathbf{x1},a1) * r(\mathbf{x1},a1) \text{ is transformational energy}$$
$$r(\mathbf{x1},a1) * f(\mathbf{x1},a1) \text{ is transformational energy}$$
Theoretical Physics considers the many variations of these variables."

Newton: "These are all different kinds of energy?"

Breton: "Yes and there are many, many more besides."

Einstein: "And my illustrious ancestor discovered one of these energies when he wrote his famous equation $E = mc^2$."

Breton: "Scalar, vectorial, or transformational?"

Einstein: "Scalar, I guess."

Breton: "Observational as extended, moving, or forcing?"

Einstein: "An atomic explosion is clearly observable."

Breton: "As a force applied through a distance, or directly as energy?"

Einstein: "What do you say?"

Breton: "All the ideas of Theoretical Physics are ideas, not physical objects or physical attributes. Energy is a derivative idea of Theoretical Physics, not like the primary attributes of extension, motion, or force which refer to something tangible. Unfortunately your illustrious ancestor, not having the well-developed language of Theoretical Physics, appears to have confused his ideas with physical objects."

Einstein, greatly alarmed: "But so much of modern physics is explained in terms of energy. Breton, you are calling into question the received wisdom of many famous names including that of my illustrious ancestor."

Breton: "Reputations aside, Theoretical Physics seeks to understand physical observations. To imagine that a spiritual idea like energy is a physical entity is a fallacy. Common enough, this kind of fallacy has even acquired a name: **reification**."

Einstein, preparing for an altercation: "We need to discuss this subject deeply and thoroughly."

Breton: "We will, I promise. If you care to pursue your illustrious ancestor's famous equation, you might want to read what I have to say in *Playing with Einstein: Reflections on E=mc²* ISBN 978-0-9844299-0-5, Library of Congress Control Number: 2013936274."

Newton, looking to refocus the conversation: "How about reciprocal vectors?"

Breton: "Let's examine the directional quotient vector
$$\mathbf{qd}(\mathbf{v}) = \mathbf{uv}/q(\mathbf{v})$$
What is $\mathbf{qd}(\mathbf{r}(\mathbf{x}1,a1))$?"

Newton: "Let $\mathbf{r}(\mathbf{x}1,a1) = r(\mathbf{x}1,a1)*\mathbf{u}(\mathbf{r}(\mathbf{x}1,a1))$. Then
$$\mathbf{qd}(\mathbf{r}(\mathbf{x}1,a1)) = \mathbf{u}(\mathbf{r}(\mathbf{x}1,a1))/r(\mathbf{x}1,a1).$$
What is this?"

Breton: "Perhaps something useful for defining spatial derivatives. But before we leave the subject of references let me ask Newton 'What is the reference for this equation of your illustrious ancestor—f=ma?'"

Newton: "The symbol *f* refers to a force; the symbol *m* refers to a mass, and the symbol *a* refers to acceleration."

Breton: "Is the force material or local?"

Newton: "Material, I suppose."

Breton: "So we might rightly write it as $\mathbf{f}(\mathbf{x})$ since it is a vector."

Newton: "Yes."

Breton: "What then is the $\mathbf{x}$, a generic object or a definite particle in the sense we have defined a particle?"

Newton: "A definite object."

Breton: "So the parts of the object are excluded?"

Newton: "Well, no. The equation applies to parts of the object as well."

Breton: "So the reference is more like the particle as we have defined it."

Newton: "Well, yes."

Breton: "And I take it the force is observed. So we can write with greater clarity
$$f(\mathbf{x1}, a1).$$
Shall we write the second symbol m($\mathbf{x1}$,a1)?"

Newton: "No. You are right to take mass as a scalar, but one that is constant."

Breton: "So it should be written m($\mathbf{x1}$). Do you agree that the third symbol should be written $\mathbf{a}(\mathbf{x1},a1)$?"

Newton: "Yes, and so we might with greater clarity
$$f(\mathbf{x1},a1) = m(\mathbf{x1}) * \mathbf{a}(\mathbf{x1},a1).$$

Breton: "To denote the force applied to the particle *x1* at observation *a1* is equal to the mass of the particle *x1* multiplied by the acceleration of the particle *x1* at observation *a1*."

Newton: "That's a lot of words which are nicely condensed into f=ma."

Breton: "With the loss of clarity which opens the way to error. Newton, have you thought what the material expression might become in a local reference? That is, 'What is $f(\mathbf{r1},a1)$?'"

Newton: "An interesting question to which I have no answer."

Breton: "The answer can be reached with Theoretical Physics, but much further up our trail up the mountain of discovery

Newton, gratefully changing the subject: "What about the set of vectors as a calculus?"

Breton: "Thank you, Newton. Even with the set of vectors as an algebra, we see the translation into Theoretical Physics carries us quickly into a discussion about the foundations of modern physics. Still, by your leave Einstein, let us continue on our path to a calculus of the set of vectors. I suspect we will encounter no few surprises, which may also impinge on modern physics. But the hour grows late, and tomorrow is another day."

With these words, the three friends arose from their Windsors, extinguished the few remaining embers in the fireplace, and left their clubhouse pensively, but with a certain anticipation for the next day.

Notes

www.ingramcontent.com/pod-product-compliance
Ingram Content Group UK Ltd.
Pitfield, Milton Keynes, MK11 3LW, UK
UKHW021329180426
11947UKWH00017B/1533